To Mary Ann

Fundamentals of Enzyme Kinetics

Athel Cornish-Bowden

Lecturer in Biochemistry,
University of Birmingham, U.K.

BUTTERWORTHS
LONDON – BOSTON
Sydney – Wellington – Durban – Toronto

The Butterworth Group

United Kingdom	**Butterworth & Co (Publishers) Ltd** London: 88 Kingsway, WC2B 6 AB
Australia	**Butterworths Pty Ltd** Sydney: 586 Pacific Highway, Chatswood, NSW 2067 Also at Melbourne, Brisbane, Adelaide and Perth
Canada	**Butterworth & Co (Canada) Ltd** Toronto: 2265 Midland Avenue, Scarborough, Ontario, M1P 4S1
New Zealand	**Butterworths of New Zealand Ltd** Wellington: T & W Young Building, 77–85 Customhouse Quay, 1, CPO Box 472
South Africa	**Butterworth & Co (South Africa) (Pty) Ltd** Durban: 152–154 Gale Street
USA	**Butterworth (Publishers) Inc** Boston: 10 Tower Office Park, Woburn, Mass. 01801

First published 1979

ISBN 0 408 10617 4

© Butterworth & Co (Publishers) Ltd, 1979

British Library Cataloguing in Publication Data

Cornish-Bowden, Athel
    Fundamentals of enzyme kinetics.
    1. Enzymes
    2. Chemical reaction, Rate of
    I. Title
    547'.758   QP601   79-40116

    ISBN 0-408-10617-4

Typeset by Scribe Design, Gillingham, Kent
Printed and bound in England by the Camelot Press, Southampton

Preface

In this book I have set out to cover enzyme kinetics in sufficient
detail for the undergraduate student with an interest in enzymology
at the end of a degree course in biochemistry. It goes rather beyond
the essential core of enzyme kinetics that all students of biochemistry
learn, but I hope that about three-quarters of the book will be suitable for non-specialist teaching. The book was originally conceived
as an appropriately abridged edition of *Principles of Enzyme Kinetics*
(published by Butterworths, 1976), but in the event I have added
as much as I have taken away, as a result of greatly expanding the
coverage of the more practical aspects of the subject.

Most of the mathematics used in enzyme kinetics is elementary,
and the apparent difficulty of the subject owes more to the necessary
profusion of symbols and equations than to any real conceptual difficulties. I have tried therefore to explain the mathematical derivations
in the book in sufficient detail to remove most of the barriers to
understanding. To test understanding, and in a few cases to carry the
theory a little beyond what is given explicitly in the text, I have
included problems at the end of each chapter, with solutions at the
end of the book.

I am grateful to the Chemistry Department of the University of
Guelph for their hospitality during the early stages of writing this
book, and for providing me with an opportunity to teach a one-
semester postgraduate course on enzyme kinetics. This experience
proved invaluable to me in the planning and writing of the book. I am
also grateful to H.B.F. Dixon, M. Gregoriou, R.H. Jackson, R.N.F.
Thorneley, C.W. Wharton and E.A. Wren, who provided many valuable suggestions for improvements or corrections.

Athel Cornish-Bowden

Contents

Principal symbols used in this book

A; a	First substrate in the forward direction; its concentration
A	(In the sequential model) Conformation characteristic of an unliganded subunit
A	Constant of integration in the Arrhenius equation
a; $\hat{a}$	Ordinate intercept of general straight line; its least-squares estimate
a_0; a_∞	Concentration of A at $t = 0$; at $t = \infty$ (i.e. at equilibrium)
A*; $a*$	Radioactively labelled A; its concentration
B; b	Second substrate in the forward direction; its concentration
B	(In the sequential model) Conformation characteristic of a subunit with ligand bound
b; $\hat{b}$	Slope of general straight line; its least-squares estimate
c	(In the symmetry model) Ratio of dissociation constants for R and T states
c	(In the sequential model) Measure of stability of AB interface
E; e_0	Enzyme; its initial (or total) concentration
EA; [EA]	Complex between E and A; its concentration (also represented by single letters such as x, as defined)
EB, EP, ES, etc.	Complexes between E and B, P, S, etc. Concentrations are expressed with square brackets or defined single letters
E_a	Arrhenius activation energy
e_i	Deviation of ith observed value from calculated value
G	Transferred group in a two-substrate two-product reaction
h	(In discussion of temperature) Planck's constant
h	(In discussion of pH) Concentration of H^+
h	(In discussion of co-operative effects) Hill coefficient
I; i	Inhibitor; its concentration

K	Equilibrium constant
K	Formation constant for transition state
k	Rate constant
K_a	Acid dissociation constant
K_{AB}; K_{BB}	(In the sequential model) Subunit interaction constant for an AB interface; same for a BB interface
K_i	Competitive-inhibition constant
K_i'	Uncompetitive-inhibition constant
K_i^A	(In multiple-substrate reactions) Inhibition constant for A (*see* p. 105)
K_m; $\hat{K}_m$; K_m^*	The Michaelis constant (concentration of substrate when $v = V/2$); its least-squares estimate; its distribution-free estimate
$\tilde{K}_m$, $\tilde{V}$	pH-corrected value of K_m, V, i.e. the values K_m, V would have if the enzyme were confined to its ideal state of protonation
K_m^A, K_m^B, etc.	Michaelis constant for A, B, etc. (observed value when inhibitors are absent and concentrations of all substrates apart from the one specified are extrapolated to infinity)
K_m^{app}, V^{app}	Apparent value of K_m, V
K_m^{exp}, V^{exp}	Expected value of K_m, V, i.e. value that would apply if non-productive binding did not occur
K_R; K_T	(In the symmetry model) Dissociation constant for loss of substrate from subunit in R conformation; same for T conformation
K_s	Dissociation constant of ES
K_s^A	Dissociation constant of EA
K_{si}	Substrate-inhibition constant
K_x	Dissociation constant of EX
K_1, K_2, etc.	Association constant for first, second, etc. ligand in sequence of binding steps
k_{+1}, k_{+2}, etc.	Rate constant for first, second, etc. steps in forward direction
k_{-1}, k_{-2}, etc.	Rate constant for first, second, etc. steps in reverse direction
k_{cat}	Catalytic constant ($= V/e_0$)
L	(In the symmetry model) Equilibrium constant between R and T states in the absence of ligand
P; p	First product; its concentration
P*; $p*$; p_0; p_∞	*See* corresponding symbols at A
P	(In the simple theory of temperature effects) Probability of reaction after molecular collision
pH	Negative logarithm of H^+ concentration

pK_a	Negative logarithm of K_a
Q; q; etc.	Second product; its concentration; *see* symbols at A
R	(In the symmetry model) Conformation that binds substrate tightly
R	Gas constant
R_x	Ratio of X concentrations for 90% and 10% saturation
S; s; etc.	Substrate; its concentration; *see* symbols at A
SS	Sum of squares of deviations
T	(In the symmetry model) Conformation that binds substrate weakly
T	Absolute temperature (in kelvins)
t	time
V; $\hat{V}$; $V*$	Maximum velocity, i.e. value of v at saturation; its least-squares estimate; its distribution-free estimate
$\widetilde{V}$; V^{app}; V^{exp}	*See* under K_m
v	Rate of reaction (at $t = 0$ if not otherwise specified)
v_0	Initial rate (subscript omitted when unambiguous)
$v*$	Rate of transfer of radioisotopic label
$\hat{v}$	Calculated rate
V^f; V^r	Value of V in forward direction; in reverse direction
X	Ligand of unspecified nature (e.g. substrate, product, activator, inhibitor)
$X^{\ddagger}$	Transition state
x, y	Concentrations of intermediates (always defined when used)
Y	Fractional saturation
Z	Collision frequency
$\Delta G^{\ddagger}$	Free energy of activation
ΔH^0	Standard enthalpy of reaction
$\Delta H^{\ddagger}$	Enthalpy of activation
$\Delta S^{\ddagger}$	Entropy of activation
τ	Relaxation time

Chapter 1

Basic principles of chemical kinetics

1.1 Order of a reaction

A chemical reaction can be classified either according to its *molecularity* or according to its *order*. The molecularity is defined by the number of molecules that are altered in the reaction. Thus, a reaction A → P is *unimolecular* (sometimes called *monomolecular*), a reaction A + B → P is *bimolecular*, and a reaction A + B + C → P is *trimolecular* (or *termolecular*). The order is a description of the kinetics of the reaction and defines how many concentration terms must be multiplied together to get an expression for the rate of reaction. Hence, in a first-order reaction, the rate is proportional to one concentration; in a second-order reaction it is proportional to two concentrations or the square of one concentration; and so on.

For a simple reaction that consists of a single step, or for each step in a complex reaction, the order is usually the same as the molecularity. However, many reactions consist of sequences of unimolecular and bimolecular steps, and the molecularity of the complete reaction need not be the same as its order. Indeed, for a complex reaction it is often not meaningful to define an order, as the rate often cannot be expressed as a product of concentration terms. As we shall see in later chapters, this is almost universal in enzyme kinetics, and even the simplest enzyme-catalysed reactions do not have simple orders. In spite of this, the concept of order is important in the understanding of enzyme kinetics, because the individual steps in enzyme-catalysed reactions nearly always do have simple orders, usually being first or second order. The binding of a substrate molecule to an enzyme molecule is a typical example of a second-order bimolecular reaction in enzyme kinetics; whereas conversion of an enzyme–substrate complex into products or into another intermediate is a typical example of a first-order unimolecular reaction.

For a first-order reaction A → P, the velocity v can be expressed as

$$v = \frac{dp}{dt} = -\frac{da}{dt} = ka = k(a_0 - p) \tag{1.1}$$

in which a and p are the concentrations at any time t of A and P respectively, and k is a *first-order rate constant*. At the start of reaction $t = 0$ and $a = a_0$. By the stoicheiometry of the reaction, every molecule of A that is consumed is converted into a molecule of P, and so the concentrations of A and P are related by the equation $a + p = a_0$. Equation 1.1 can readily be integrated by separating the two variables p and t, i.e. by bringing all terms in p on to the left-hand side and all terms in t on to the right-hand side:

$$\int \frac{dp}{a_0 - p} = \int k \, dt$$

Therefore

$$-\ln (a_0 - p) \doteq kt + \alpha$$

in which α is a constant of integration. This can be evaluated by allowing for the fact that there is no product at the start of the reaction, i.e. $p = 0$ when $t = 0$. Then, $\alpha = -\ln (a_0)$, and so

$$\ln [(a_0 - p)/a_0] = -kt$$

Taking exponentials of both sides we have

$$(a_0 - p)/a_0 = \exp (-kt)$$

which can be rearranged to give

$$p = a_0 [1 - \exp (-kt)] \tag{1.2}$$

It is important to note that the constant of integration, α, was included in this derivation, evaluated and found to be non-zero. Constants of integration must always be included and calculated when kinetic equations are integrated; they are rarely found to be zero.

The commonest type of bimolecular reaction is one of the form $A + B \rightleftharpoons P + Q$, in which two different types of molecule, A and B, react to give products. In this case the rate is likely to be given by a second-order expression of the form

$$v = dp/dt = kab = k(a_0 - p)(b_0 - p) \tag{1.3}$$

in which k is now a *second-order rate constant*. (Notice that conventional symbolism does not, unfortunately, indicate the order of a rate constant.) Again, integration is readily achieved by separating the two variables p and t:

$$\int \frac{dp}{(a_0 - p)(b_0 - p)} = \int k \, dt$$

For readers with limited mathematical experience, the simplest and most reliable method for integrating the left-hand side of this equation is to look it up in standard tables of integrals. It may also be done by multiplying both sides of the equation by $(b_0 - a_0)$ and separating the left-hand side into two simple integrals:

$$\int \frac{dp}{a_0 - p} - \int \frac{dp}{b_0 - p} = \int (b_0 - a_0) k \, dt$$

Hence

$$-\ln (a_0 - p) + \ln (b_0 - p) = (b_0 - a_0)kt + \alpha$$

Putting $p = 0$ when $t = 0$, we find $\alpha = \ln (b_0/a_0)$, and so

$$\ln \left[\frac{a_0 (b_0 - p)}{b_0 (a_0 - p)} \right] = (b_0 - a_0)kt$$

or

$$\frac{a_0 (b_0 - p)}{b_0 (a_0 - p)} = \exp [(b_0 - a_0)kt] \qquad (1.4)$$

The following special case of this result is of interest: if a_0 is very small compared with b_0, then p must also be very small compared with b_0 at all times, because p can never exceed a_0 on account of the stoicheiometry of the reaction. So $(b_0 - a_0)$ and $(b_0 - p)$ can both be written with good accuracy as b_0 and equation 1.4 simplifies to

$$p = a_0 [1 - \exp (-kb_0 t)]$$

This is of exactly the same form as equation 1.2, the equation for a first-order reaction. This type of reaction is known as a *pseudo-first-order* reaction, and kb_0 is a *pseudo-first-order rate constant*. The situation occurs naturally when one of the reactants is the solvent, as in most hydrolysis reactions, but it is also advantageous to set up pseudo-first-order conditions deliberately, in order to simplify evaluation of the rate constant, as I shall discuss in Section 1.5.

Trimolecular reactions, such as $A + B + C \rightarrow P + \ldots$, do not usually consist of a single trimolecular step, and consequently they are not usually third order. Instead the reaction is likely to consist of two or more *elementary steps*, such as

$A + B \rightarrow X$

and

$X + C \rightarrow P$

If one step in such a reaction is much slower than the others, the rate of the complete reaction is equal to the rate of the slow step, which is accordingly known as the *rate-determining* (or *rate-limiting*) step. If there is no clearly defined rate-determining step, the rate equation is likely to be complex and to have no clear order. Some trimolecular reactions do display third-order kinetics, however, with $v = kabc$, where k is now a *third-order rate constant*, but it is *not* necessary to assume a three-body collision (which is inherently very unlikely) to account for third-order kinetics. Instead, we can assume a two-step mechanism, as above but with the first step rapidly reversible, so that the concentration of X is given by $x = Kab$, where K is the equilibrium constant for binding of A to B, i.e. the association constant of X. The rate of reaction is then the rate of the slow second step:

$$v = k'xc = k'Kabc$$

where k' is the second-order rate constant for the second step. Hence the observed third-order rate constant is actually the product of a second-order rate constant and an equilibrium constant.

Some reactions are observed to be *zero order*, that is, the rate is found to be constant, independent of the concentration of reactant. If a reaction is zero order with respect to only one reactant, this may simply mean that the reactant enters the reaction after the rate-determining step. However, some reactions are zero order overall, that is, independent of all reactant concentrations. Such reactions are invariably catalysed reactions and occur if every reactant is present in such large excess that the full potential of the catalyst is realized. Zero-order kinetics are common in enzyme-catalysed reactions as the limit at very high reactant concentrations.

1.2 Determination of the order of a reaction

The simplest means of determining the order of a reaction is to measure the rate at different concentrations of the reactants. Then a plot of log (rate) against log (concentration) gives a straight line with slope equal to the order. If all of the reactant concentrations are altered in a constant ratio, the slope of the line is the overall order. It is useful to know the order with respect to each reactant, however, and this can be found by altering the concentration of each reactant separately, keeping the other concentrations constant. Then the slope of the line will be equal to the order with respect to the variable reactant. For example, if the reaction is second order in A and first order in B,

$$v = ka^2 b$$

then

$$\log v = \log k + 2 \log a + \log b$$

Hence a plot of log v against log a (with b held constant) will have a slope of 2, and a plot of log v against log b (with a held constant) will have a slope of 1. These plots are illustrated in *Figure 1.1*. It is important to realize that if the rates are determined from the slopes of the progress curve (i.e. a plot of concentration against time), the concentrations of all of the reactants will change. Therefore, if valid results are to be obtained, either the initial concentrations of the

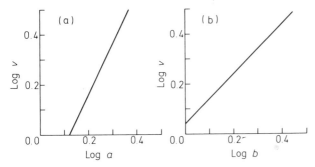

Figure 1.1 Determination of the order of reaction: the lines are drawn for a reaction that is second order in A and first order in B, so the slopes of the plots are 2 and 1 respectively

reactants must be in stoicheiometric ratio, in which event the over-all order will be found, or (more usually) the 'constant' reactants must be in large excess at the start of the reaction, so that the changes in their concentrations are insignificant. If neither of these alternatives is possible or convenient, the rates must be obtained from a set of measurements of the slope at zero time, i.e. of initial rates. This method is usually preferable for kinetic measurements of enzyme-catalysed reactions, because the progress curves of enzyme-catalysed reactions often do not rigorously obey simple rate equations for extended periods of time. The progress curve of an enzyme-catalysed reaction often requires a more complicated equation than the integrated form of the rate equation derived for the initial rate, because of progressive loss of enzyme activity, inhibition by products, and other effects.

1.3 Dimensions of rate constants

Dimensional analysis provides one of the simplest and most versatile techniques for detecting algebraic mistakes and checking results available in enzyme kinetics. It depends on the existence of a few simple rules governing the permissible ways of combining quantities of different dimensions, and on the fact that algebraic errors frequently result in dimensionally inconsistent expressions. Concentrations can be expressed in M (or mol l^{-1}), and reaction rates in

$M\ s^{-1}$. In an expression such as $v = ka$, therefore, the rate constant k must be expressed in s^{-1} if the left- and right-hand sides of the equation are to have the same dimensions. All first-order rate constants have the dimension $(time)^{-1}$, and by a similar argument second-order rate constants have the dimensions $(concentration)^{-1}(time)^{-1}$, third-order rate constants $(concentration)^{-2}(time)^{-1}$, and zero-order rate constants $(concentration)(time)^{-1}$.

Knowledge of the dimensions of rate constants allows the correctness of derived equations to be checked very easily: the left- and right-hand sides of any equation (or inequality) must have the same dimensions, and all of the terms in a summation must have the same dimensions. For example, if $(1 + t)$ occurs in an equation, where t has the dimension (time), then either the equation is incorrect, or the '1' is a time that happens to have a numerical value of 1. Quantities of different dimensions can be multiplied or divided, but must not be added or subtracted. Thus, if k_1 is a first-order rate constant and k_2 is a second-order rate constant, a statement such as $k_1 \gg k_2$ is meaningless, just as $5\ g \gg 25°C$ is meaningless. However, a pseudo-first-order rate constant such as $k_2 a$ has the dimensions $(concentration)^{-1}(time)^{-1}(concentration)$, i.e. $(time)^{-1}$; it therefore has the dimensions of a first-order rate constant, and *can* be compared with other first-order rate constants.

Another major principle of dimensional analysis is that one must not use a dimensioned quantity as an exponent or take its logarithm. For example, $\exp(-kt)$ is permissible, provided that k is a first-order rate constant, but $\exp(-t)$ is not. There is an apparent exception to

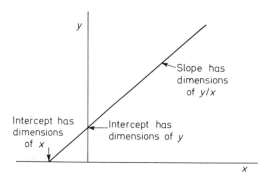

Figure 1.2 Dimensional analysis as applied to graphs

the rule that it is not permissible to take the logarithm of a dimensioned quantity, in that it is often convenient to take the logarithm of what appears to be a concentration, e.g. in the definition of pH as $-\log[H^+]$. The explanation of this is that the definition is not strictly accurate and to be dimensionally correct one should define pH as $-\log\{[H^+]/[H^+]^0\}$, where $[H^+]^0$ is the value of $[H^+]$ in the standard state, i.e. at pH = 0. As $[H^+]^0$ has a numerical value of 1 it is usually omitted from the definition. Whenever one takes the logarithm of a dimensioned quantity in this way, a standard state is implied whether stated explicitly or not.

Dimensional analysis is particularly useful as an aid to remembering the slopes and intercepts of commonly used plots, and the rules are very simple: any intercept must have the same dimensions as whatever variable is plotted along the corresponding axis, and a slope must have the dimensions of the ordinate (y) divided by those of the abscissa (x). These rules are illustrated in *Figure 1.2*.

1.4 Reversible reactions

Many reactions are readily reversible, and the back reaction must be allowed for in the rate equation:

$$A \underset{k_{-1}}{\overset{k_{+1}}{\rightleftarrows}} P$$

$$a_0 - p \qquad p$$

In this case,

$$v = \frac{dp}{dt} = k_{+1}(a_0 - p) - k_{-1}p = k_{+1}a_0 - (k_{+1} + k_{-1})p \qquad (1.5)$$

This differential equation is of exactly the same form as equation 1.1, and can be solved in the same way:

$$\int \frac{dp}{k_{+1}a_0 - (k_{+1} + k_{-1})p} = \int dt$$

Therefore

$$\frac{\ln [k_{+1}a_0 - (k_{+1} + k_{-1})p]}{-(k_{+1} + k_{-1})} = t + \alpha$$

Setting $p = 0$ when $t = 0$ gives $\alpha = -\ln (k_{+1}a_0)/(k_{+1} + k_{-1})$, and so

$$\ln \left(\frac{k_{+1}a_0 - (k_{+1} + k_{-1})p}{k_{+1}a_0} \right) = -(k_{+1} + k_{-1})t$$

Taking exponentials of both sides, we have

$$\frac{k_{+1}a_0 - (k_{+1} + k_{-1})p}{k_{+1}a_0} = \exp[-(k_{+1} + k_{-1})t]$$

which can be rearranged to give

$$p = \frac{k_{+1}a_0\{1 - \exp\left[-(k_{+1} + k_{-1})t\right]\}}{k_{+1} + k_{-1}} = p_\infty\{1 - \exp\left[-(k_{+1} + k_{-1})t\right]\}$$

$$(1.6)$$

where $p_\infty = k_{+1}a_0/(k_{+1} + k_{-1})$ is the value of p after infinite time, i.e. at equilibrium.

1.5 Determination of first-order rate constants

Very many reactions are first-order in each reactant, and in these cases it is often possible to carry out the reaction under pseudo-first-order conditions overall by keeping every reactant except one in large excess. Thus, in many practical situations, the problem of determining a rate constant can be reduced to the problem of determining a first-order rate constant. We have seen (equation 1.2) that for a simple first-order reaction,

$$p = a_0[1 - \exp(-kt)]$$

and in the more general case of a reversible reaction (equation 1.6),

$$p = p_\infty\{1 - \exp\left[-(k_{+1} + k_{-1})t\right]\}$$

So

$$p_\infty - p = p_\infty \exp\left[-(k_{+1} + k_{-1})t\right] \qquad (1.7)$$

Therefore,

$$\ln(p_\infty - p) = \ln p_\infty - (k_{+1} + k_{-1})t$$

or, more conveniently,

$$\log(p_\infty - p) = \log p_\infty - [(k_{+1} + k_{-1})t/2.303]$$

Thus, a plot of $\log(p_\infty - p)$ against t gives a straight line of slope $-(k_{+1} + k_{-1})/2.303$.

Guggenheim (1926) pointed out a major objection to this plot, in that it depends very heavily on an accurate value of p_∞. In the general case where $p_\infty \neq a_0$, an accurate value of p_∞ is difficult to obtain, and even in the special case of an irreversible reaction when $p_\infty = a_0$, the instantaneous concentration of A at zero time may be difficult to measure accurately. Guggenheim suggested measuring two sets of values, p_i and p'_i, at times t_i and t'_i, such that every $t'_i = t_i + \tau$, where τ is a constant. Then, from equation 1.7,

$$p_\infty - p_i = p_\infty \exp\left[-(k_{+1} + k_{-1})t_i\right]$$

$$p_\infty - p'_i = p_\infty \exp\left[-(k_{+1} + k_{-1})(t_i + \tau)\right]$$

By subtraction,

$$p'_i - p_i = p_\infty\{1 - \exp\left[-(k_{+1} + k_{-1})\tau\right]\} \exp\left[-(k_{+1} + k_{-1})t_i\right]$$

Taking logarithms,

$$\ln (p_i' - p_i) = \ln p_\infty + \ln \{1 - \exp [k_{+1} + k_{-1})\tau]\} - (k_{+1} + k_{-1})t_i$$

or

$$\log (p_i' - p_i) = \text{constant} - (k_{+1} + k_{-1})t_i/2.303$$

So a plot of $\log (p_i' - p_i)$ against t_i gives a straight line of slope $-(k_{+1} + k_{-1})/2.303$, as illustrated in *Figure 1.3*. This is known as a *Guggenheim plot*, and it has the major advantage that it does not

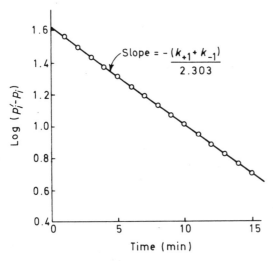

Figure 1.3 Determination of a first-order rate constant by means of a Guggenheim plot: *p* is the concentration of product at time *t* and *p'* is the concentration of product at time $(t + \tau)$, where τ is a constant

require an estimate of p_∞. As k_{+1}/k_{-1} is equal to the equilibrium constant, which can be estimated independently, the values of the individual rate constants k_{+1} and k_{-1} can be calculated from the two combinations.

1.6 Influence of temperature on rate constants

From the earliest studies of reaction velocities, it has been evident that they are profoundly influenced by temperature. The most elementary consequence of this is that the temperature must always be controlled if meaningful results are to be obtained from kinetic experiments. However, with care, one can use temperature much more positively and, by carrying out measurements at several temperatures, one can deduce important information about reaction mechanisms.

The studies of van't Hoff (1884) and Arrhenius (1889) form the starting point for all modern theories of the temperature dependence of rate constants. Harcourt (1867) had earlier noted that the rates of many reactions approximately doubled for each $10\,°C$ rise in temperature, but van't Hoff and Arrhenius attempted to find a more exact relationship by comparing kinetic observations with the known properties of equilibrium constants. Any equilibrium constant K varies with the absolute temperature T in accordance with the van't Hoff equation,

$$\frac{d\ln K}{dT} = \frac{\Delta H^0}{RT^2}$$

where R is the gas constant and ΔH^0 is the standard enthalpy change in the reaction. But K can be regarded as the ratio k_+/k_- of the rate constants k_+ and k_- for the forward and reverse reactions (because the net rate of any reaction is zero at equilibrium). So we can write

$$\frac{d\ln (k_+/k_-)}{dT} = \frac{d\ln k_+}{dT} - \frac{d\ln k_-}{dT} = \frac{\Delta H^0}{RT^2}$$

This equation can be partitioned as follows to give separate expressions for k_+ and k_-:

$$\frac{d\ln k_+}{dT} = \frac{\Delta H^0_+}{RT^2} + \lambda$$

$$\frac{d\ln k_-}{dT} = \frac{\Delta H^0_-}{RT^2} + \lambda$$

where λ is a quantity about which nothing can be said *a priori* except that it must be the same in both equations (because otherwise it would not vanish when one equation is subtracted from the other). Thus far this derivation follows from thermodynamic considerations and involves no assumptions. However, it proved difficult or impossible to show experimentally that the term λ in these equations was necessary. So Arrhenius postulated that its value was in fact zero, and that the temperature dependence of any rate constant k could be expressed by an equation of the form

$$\frac{d\ln k}{dT} = \frac{E_a}{RT^2} \tag{1.8}$$

where E_a is the *activation energy* and corresponds to the standard enthalpy of reaction ΔH^0 in the van't Hoff equation. Integration with respect to T gives

$$\ln k = \ln A - (E_a/RT) \tag{1.9}$$

where $\ln A$ is a constant of integration. This form of the Arrhenius equation is the most convenient for graphical purposes, as it shows that a plot of ln k against $1/T$ is a straight line of slope $-E_a/R$, or, if log k is plotted against $1/T$, the slope is $-E_a/2.303R$. This plot, which is illustrated in *Figure 1.4*, is known as an *Arrhenius plot*, and provides a simple method of evaluating E_a.

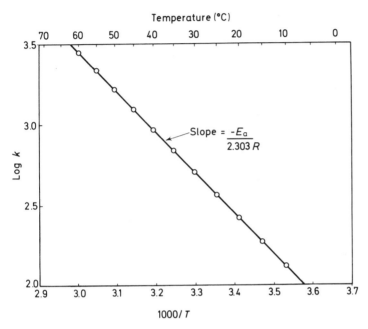

Figure 1.4 Arrhenius plot

After taking exponentials, equation 1.9 may be rearranged to give

$$k = A \exp (-E_a/RT)$$

According to Boltzmann's theory of the distribution of energies amongst molecules, the number of molecules in a mixture that have energies in excess of E_a is proportional to exp $(-E_a/RT)$. We can therefore interpret the Arrhenius equation to mean that molecules can take part in a reaction only if their energy exceeds some threshold value, the activation energy. In this interpretation, the constant A ought to be equal to the frequency of collisions, Z, at least for bimolecular reactions. For some simple reactions in the gas phase, such as the decomposition of hydrogen iodide, A is indeed equal to Z, but in general it is necessary to introduce a factor P,

$$k = PZ \exp (-E_a/RT)$$

and to assume that, in addition to colliding with sufficient energy, molecules must also be correctly oriented if they are to react. The factor P is then taken to be a measure of the probability that the correct orientation will be adopted spontaneously. This equation is now reasonably in accordance with modern theories of reaction rates, but for most purposes it is profitable to approach the same result from a different point of view, known as the transition-state theory, which is discussed in the next section.

1.7 Transition-state theory

The transition-state theory is derived largely from the work of Eyring (1935), and is so called because it attempts to relate the rates of chemical reactions to the thermodynamic properties of a particular high-energy state of the reacting molecules, known as the *transition state*, or *activated complex*. As a reacting system proceeds along a notional 'reaction co-ordinate', it must pass through a continuum of energy states, as illustrated in *Figure 1.5*, and at some

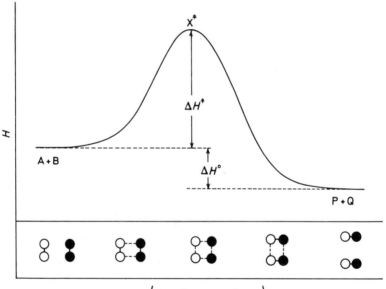

Figure 1.5 'Reaction profile' according to the transition-state theory. The diagrams along the abscissa indicate the meaning of the 'reaction co-ordinate' for a simple bimolecular reaction

stage it must surpass a state of maximum energy. This maximum energy state is the transition state, and should be clearly distinguished from an *intermediate*, which represents not a maximum but a meta-stable minimum on the reaction profile. A bimolecular reaction can be represented as

$$A + B \xrightleftharpoons{K^\ddagger} X^\ddagger \longrightarrow P + Q$$

where $X^\ddagger$ is the transition state. Its concentration is assumed to be governed by the laws of thermodynamics, so that $[X^\ddagger] = K^\ddagger[A][B]$, where $K^\ddagger$ is given by

$$\Delta G^\ddagger = -RT \ln K^\ddagger = \Delta H^\ddagger - T\Delta S^\ddagger$$

where $\Delta G^\ddagger$, $\Delta H^\ddagger$ and $\Delta S^\ddagger$ are the free energy, enthalpy and entropy of formation, respectively, of the transition state from the reactants. The concentration of $X^\ddagger$ is therefore given by

$$[X^\ddagger] = [A][B] \exp(\Delta S^\ddagger/R) \exp(-\Delta H^\ddagger/RT)$$

As written, this equation, like any true thermodynamic equation, contains no information about time. To introduce time, we require quantum-mechanical principles that are beyond the scope of this book (see, for example, Laidler, 1965), and the rate constant for the breakdown of $X^\ddagger$ can be shown to be RT/Nh, where R is the gas constant, N is Avogadro's number and h is Planck's constant. (The numerical value of RT/Nh is about 6.25×10^{12} s^{-1} at 300 K.) Therefore, the second-order rate constant for the complete reaction is

$$k = \frac{RT}{Nh} \exp(\Delta S^\ddagger/R) \exp(-\Delta H^\ddagger/RT) \qquad (1.10)$$

Taking logarithms, we obtain

$$\ln k = \ln(RT/Nh) + (\Delta S^\ddagger/R) - (\Delta H^\ddagger/RT)$$

and differentiating,

$$\frac{d \ln k}{dT} = (\Delta H^\ddagger + RT)/RT^2$$

Comparing this equation with the Arrhenius equation (equation 1.8), one can see that the activation energy E_a is not equal to $\Delta H^\ddagger$, but to $\Delta H^\ddagger + RT$. Moreover, E_a is not strictly independent of temperature, so the Arrhenius plot ought to be curved. However, the expected curvature is so slight that one would not normally expect to detect it, and the variation in k that results from the factor T in equation 1.10 is trivial in comparison with variation in the exponential term.

As both A and E_a in equation 1.9 can readily be determined in practice from an Arrhenius plot, both $\Delta H^\ddagger$ and $\Delta S^\ddagger$ can be calculated, from

$$\Delta H^\ddagger = E_a - RT$$

$$\Delta S^\ddagger = R \ln(ANh/RT) - R$$

The enthalpy and entropy of activation of a chemical reaction provide valuable information about the nature of the transition state, and hence about the reaction mechanism. A large enthalpy of activation indicates that a large amount of stretching, squeezing or even breaking of chemical bonds is necessary for the formation of the transition state.

The entropy of activation gives a measure of the inherent probability of the transition state, apart from energetic considerations. If $\Delta S^{\ddagger}$ is large and negative, the formation of the transition state requires the reacting molecules to adopt precise conformations and approach one another at a precise angle. As molecules vary widely in their conformational stability, that is, their rigidity, and in their complexity, one might expect that the values of $\Delta S^{\ddagger}$ would vary widely between different reactions. This does, in fact, occur. The molecules that are important in metabolic processes are mostly large and flexible, and so uncatalysed reactions between them are inherently unlikely, i.e. $-\Delta S^{\ddagger}$ is usually large.

Equation 1.10 shows that a catalyst can increase the rate of a reaction either by reducing $-\Delta S^{\ddagger}$ or by reducing $\Delta H^{\ddagger}$, or both. It is likely that both effects are important in enzymic catalysis, although in most cases it is not possible to obtain definite evidence of this because the uncatalysed reactions are too slow for $-\Delta S^{\ddagger}$ and $\Delta H^{\ddagger}$ to be measured.

Problems

1.1 The following data were obtained for the rate of a reaction with stoicheiometry A + B → P at various concentrations of A and B:

[A] (mM)	10	20	50	100	10	20	50	100
[B] (mM)	10	10	10	10	20	20	20	20
v (μmol l^{-1} s^{-1})	0.6	1.0	1.4	1.9	1.3	2.0	2.9	3.9

[A] (mM)	10	20	50	100	10	20	50	100
[B] (mM)	50	50	50	50	100	100	100	100
v (μmol l^{-1} s^{-1})	3.2	4.4	7.3	9.8	6.3	8.9	14.4	20.3

Determine the order with respect to A and B. Suggest an explanation for the order with respect to A.

1.2 Check the following statements for dimensional consistency, assuming that t represents time (s), v and V represent rates (M s^{-1} or mol l^{-1} s^{-1}), and a, p, s and K_m represent concentrations (M):

(a) In a plot of v against v/s, the slope is $-1/K_m$ and the intercept on the v/s axis is K_m/V.

(b) In a bimolecular reaction $2A \rightarrow P$, with rate constant k, the concentration of P at time t is given by $p = a_0^2 kt/(1 + 2a_0 kt)$.

(c) A plot of $t/\ln (s_0/s)$ against $(s_0 - s)/\ln (s_0/s)$ for an enzyme-catalysed reaction gives a straight line of slope $1/V$ and intercept V/K_m on the ordinate.

1.3 Many reactions display an approximate doubling of rate when the temperature is raised from 25 °C to 35 °C. What does this imply about their enthalpies of activation? ($R = 8.31$ J mol^{-1} K^{-1}, 0 °C = 273 K, ln 2 = 0.693.)

1.4 In the derivation of the Arrhenius equation (Section 1.6), a term λ was introduced and subsequently assumed to be zero. In the light of the transition-state theory (Section 1.7), what would you expect the value of λ to be at 300 K (27 °C)?

For additional problems in the use of dimensional analysis, see Problems 7.2 (p. 145) and 10.4 (p. 211).

Chapter 2

Introduction to enzyme kinetics

2.1 Early studies: the idea of an enzyme–substrate complex

The rates of enzyme-catalysed reactions were first studied in the
latter part of the nineteenth century. At that time, no enzyme was
available in a pure form, methods of assay were primitive, and the
use of buffers to control pH had not been introduced. Moreover, it
was customary to follow the course of the reaction over a period of
time, in contrast to the more usual modern practice of measuring
initial rates at various different initial substrate concentrations, which
gives results that are easier to interpret.

Most of the early studies were concerned with enzymes from fermentation, particularly invertase, which catalyses the hydrolysis of
sucrose:

sucrose + water → glucose + fructose

O'Sullivan and Tompson (1890) studied this reaction, and made a
number of important discoveries: they found that the reaction was
highly dependent on the acidity of the mixture and that, provided
that 'the acidity was in the most favourable proportion', the rate was
proportional to the amount of enzyme. The rate decreased as the
substrate was consumed, and seemed to be proportional to the
sucrose concentration, though there were slight deviations from the
expected curve. At low temperatures, the enzyme showed an approximate doubling of rate for an increase in temperature of 10 °C. However, unlike most ordinary chemical reactions, the invertase-catalysed
reaction displayed an apparent *optimum temperature*, above which
the rate fell rapidly to zero. Invertase proved to be a true catalyst,
as it was not destroyed or altered in the reaction (except at high temperatures), and a sample was still active after catalysing the hydrolysis of 100 000 times its weight of sucrose. Finally, the thermal
stability of the enzyme was very much greater in the presence of its
substrate than in its absence: 'Invertase when in the presence of cane
sugar [i.e. sucrose] will stand without injury a temperature fully
25 °C greater than in its absence. This is a very striking fact, and, as

far as we can see, there is only one explanation of it, namely, the invertase enters into combination with the sugar.' Wurtz (1880) had reached a similar conclusion previously: while studying the papain-catalysed hydrolysis of fibrin, he observed a precipitate that he suggested might be a papain–fibrin compound that acted as an inter-mediate in the hydrolysis.

Brown (1892) placed the idea of an *enzyme–substrate complex* in a purely kinetic context. In common with a number of other workers, he found that the rates of enzyme-catalysed reactions deviated from second-order kinetics. Initially, he showed that the rate of hydrolysis of sucrose in fermentation by live yeast appeared to be independent of the sucrose concentration. The conflict between Brown's results with live yeast and those of O'Sullivan and Tompson with isolated invertase was not at first regarded as serious, because catalysis by isolated enzymes was regarded as fundamentally different from fermentation by living organisms. But Buchner's (1897) discovery that a cell-free (i.e. non-living) extract of yeast could catalyse alcoholic fermentation prompted Brown (1902) to re-examine his earlier results. After confirming that they were correct, he showed that similar results could be obtained with purified inver-tase. He suggested that the enzyme–substrate complex mechanism placed a limit on the rate that could be achieved. Provided that the complex existed for a brief instant of time before breaking down to products, then a maximum rate would be reached when the sub-strate concentration was high enough to convert all of the enzyme into complex, according to the law of mass action. At lower concen-trations of substrate, the rate at which complex was formed would become significant, and so the rate of hydrolysis would be depen-dent on substrate concentration.

2.2 Michaelis–Menten equation

Henri (1902, 1903) criticized Brown's model of enzyme action on the grounds that it assumed a fixed lifetime for the enzyme–substrate complex between its abrupt creation and decay. He proposed instead a mechanism that was conceptually very similar to Brown's but which was proposed in more precise mathematical and chemical terms, with an equilibrium between the free enzyme and the enzyme–substrate and enzyme–product complexes.

Now although Brown and Henri reached essentially correct con-clusions, they did so on the basis of experiments that were open to serious objections. O'Sullivan and Tompson experienced great diffi-culty in obtaining coherent results until they realized the importance of acid concentration. Brown prepared the enzyme in a different way and found the addition of acid to be unnecessary (presumably his solutions were weakly buffered by the natural components of the yeast), and Henri did not discuss the problem. Apart from O'Sullivan and Tompson, the early investigators of invertase made no allowance for the mutarotation of the glucose produced in the reaction, although this undoubtedly affected the results because they used polarimetric methods for following the reaction.

With the introduction of the concept of hydrogen-ion concentration, expressed by the logarithmic scale of pH (Sørensen, 1909), Michaelis and Menten (1913) realized the necessity for carrying out definitive experiments with invertase. They controlled the pH of the reaction by the use of acetate buffers, they allowed for the mutarotation of the product and they measured *initial rates* of the reaction at different substrate concentrations. If initial rates are used, the reverse reaction, inhibition by products, progressive inactivation of the enzyme and other complicating features can be avoided, and a much simpler rate equation can be used. In spite of these refinements, Michaelis and Menten obtained results in good agreement with those of Henri, and they proposed a mechanism essentially the same as that developed by him:

$$E + S \rightleftharpoons ES \rightarrow E + P$$

Like Henri, they assumed that the reversible first step was fast enough for it to be represented by an equilibrium constant, $K_s = es/x$, where x is the concentration of intermediate, ES, so that $x = es/K_s$. The instantaneous concentrations of free enzyme and substrate, e and s respectively, are not directly measurable, however, and so they must be expressed in terms of the initial, measured, concentrations, e_0 and s_0, using the stoicheiometric relationships

$$e_0 = e + x$$

and

$$s_0 = s + x$$

From the first of these, x cannot be greater than e_0, and so, provided that s_0 is much larger than e_0, it must also be much larger than x. So $s = s_0$ with good accuracy. Then the expression for x becomes

$$x = (e_0 - x)s/K_s$$

which can be rearranged to give

$$x = \frac{e_0}{(K_s/s) + 1}$$

The second step in the reaction, ES $\rightarrow$ E + P, is a simple first-order reaction, with a rate constant that may be defined as k_{+2}, so that

$$v = k_{+2}x = \frac{k_{+2}e_0}{(K_s/s) + 1} = \frac{k_{+2}e_0 s}{K_s + s} \tag{2.1}$$

Michaelis and Menten showed that this theory, and equation 2.1, could account accurately for their results with invertase. Because of the definitive nature of their experiments, which have served as a standard for most subsequent enzyme-kinetic measurements, Michaelis and Menten are regarded as the founders of modern enzymology, and equation 2.1 (in its modern form, equation 2.7, below) is generally known as the *Michaelis–Menten equation*, though an equivalent equation had been derived earlier by Henri (1902, 1903).

At about the same time, Van Slyke and Cullen (1914) obtained similar results with the enzyme urease. They assumed a similar mechanism, with the important difference that the first step was assumed to be irreversible:

$$E + S \xrightarrow{k_{+1}} ES \xrightarrow{k_{+2}} E + P$$
$$e_0 - x \quad s \qquad\quad x \qquad\qquad p$$

In this case there are no reversible reactions and so there can be no question of representing x by an equilibrium constant; instead, we have

$$dx/dt = k_{+1}(e_0 - x)s - k_{+2}x$$

Van Slyke and Cullen implicitly assumed that the intermediate concentration was constant, i.e. $dx/dt = 0$, and so

$$k_{+1}(e_0 - x)s - k_{+2}x = 0$$

which may be rearranged to give the following expression for x:

$$x = \frac{k_{+1}e_0 s}{k_{+2} + k_{+1}s}.$$

Substituting this into the rate equation $v = k_{+2}x$, we have

$$v = k_{+2}x = \frac{k_{+1}k_{+2}e_0 s}{k_{+2} + k_{+1}s} = \frac{k_{+2}e_0 s}{(k_{+2}/k_{+1}) + s} \tag{2.2}$$

This equation is of the same form as equation 2.1, with k_{+2}/k_{+1} replacing K_s, and is empirically indistinguishable from it.

At about the same time as these developments were taking place in the understanding of enzyme catalysis, Langmuir (1916, 1918) was reaching similar conclusions in a study of the adsorption of gases on to solids. His treatment was much more general, but the case that he referred to as *simple adsorption* corresponds closely to the type of binding assumed by Henri and by Michaelis and Menten. Langmuir recognized the similarity between solid surfaces and enzymes, although he imagined the whole surface of an enzyme to be 'active', rather than limited areas or *active sites*. Hitchcock (1926) pointed out the similarity between the equations for the binding of ligands to solid surfaces and to proteins, and the logical process was completed by Lineweaver and Burk (1934), who extended Hitchcock's ideas to catalysis.

2.3 Steady-state treatment

The formulation of Michaelis and Menten, which treats the first step of enzyme catalysis as an equilibrium, and that of Van Slyke and Cullen, which treats it as irreversible, both make unwarranted and unnecessary assumptions about the magnitudes of the rate constants. As we have seen, both formulations lead to the same form of the rate equation, and Briggs and Haldane (1925) examined a generalized mechanism that includes both special cases:

$$E + S \underset{k_{-1}}{\overset{k_{+1}}{\rightleftharpoons}} ES \xrightarrow{k_{+2}} E + P$$

$$\begin{array}{cccc} e_0 - x & s & x & p \end{array}$$

In this case,

$$dx/dt = k_{+1}(e_0 - x)s - k_{-1}x - k_{+2}x \tag{2.3}$$

Briggs and Haldane argued that a steady state would be reached in which the concentration of intermediate was constant, i.e. $dx/dt = 0$; then

$$k_{+1}(e_0 - x)s - k_{-1}x - k_{+2}x = 0 \tag{2.4}$$

Collecting terms in x and rearranging, we have the following expression for the steady-state value of x:

$$x = \frac{k_{+1}e_0 s}{k_{-1} + k_{+2} + k_{+1}s} \tag{2.5}$$

As before, the rate is given by $k_{+2}x$, i.e.

$$v = k_{+2}x = \frac{k_{+1}k_{+2}e_0 s}{k_{-1} + k_{+2} + k_{+1}s} = \frac{k_{+2}e_0 s}{\dfrac{k_{-1} + k_{+2}}{k_{+1}} + s} \tag{2.6}$$

This equation can be written in the more general form

$$v = Vs/(K_m + s) \tag{2.7}$$

which is the standard form of the *Michaelis–Menten equation*, with two constants, V, known as the *maximum velocity*, and K_m, known as the *Michaelis constant*. Comparison with equation 2.6 shows that, provided the mechanism defined at the beginning of this section applies, V has the value $k_{+2}e_0$ and K_m has the value $(k_{-1} + k_{+2})/k_{+1}$. However, equation 2.7 applies to many mechanisms more complex than the simple two-step Michaelis–Menten mechanism, and in general one cannot assume that V is equivalent to $k_{+2}e_0$ or that K_m is equivalent to $(k_{-1} + k_{+2})/k_{+1}$.

To avoid confusion with v, V is usually spoken aloud as 'vee-max', and is sometimes printed as V_{max} or V_m. Of these, V_{max} is harmless but is not recommended by the Commission on Biochemical Nomenclature (1973); but V_m is definitely to be avoided because it misleadingly suggests that the subscript 'm' corresponds to that in K_m. (I have occasionally seen K_{max} in answers to examination questions!) In fact, the 'm' in K_m stands for Michaelis and it was the former and more logical custom to write it as K_M.

V is not a fundamental property of an enzyme, because it depends on the enzyme concentration. At least in the early stages of investigation of an enzyme, the true enzyme molarity is usually unknown; but if e_0 can be measured in meaningful units it is advantageous to define a more fundamental quantity $k_{cat} = V/e_0$, known as the *catalytic constant* or *turnover number*. The latter name derives from the fact that k_{cat} is a reciprocal time and defines the number of catalytic processes (or 'turnovers') the enzyme can catalyse in unit time. For the simple Michaelis–Menten mechanism k_{cat} is identical with k_{+2}, but this is not necessarily true in more complex cases where the Michaelis–Menten equation applies.

For enzymes whose molar concentration cannot be measured, either because the enzyme has not been purified or because its molecular weight is unknown, it is often convenient to define a unit of catalytic activity. The traditional 'unit' of enzymologists is the amount of enzyme that can catalyse the transformation of 1 μmol of substrate into products in 1 min under standard conditions. This unit is still in common use, because the corresponding SI unit, the 'katal', abbreviation kat, has yet to gain popularity with enzymologists. Their lack of enthusiasm derives in part from the fact that 1 kat is an inconveniently large amount of catalytic activity, as it is the amount sufficient to catalyse the transformation of 1 mol of substrate into products in 1 s under standard conditions: this may be compared with a typical enzyme activity of 1 unit/ml in a cell extract, or about 20 μkat/l. Similar, indeed more severe, objections to the farad as the unit of capacitance have not prevented it from becoming generally accepted by electrical engineers as the standard unit, though they more often use submultiples of it. Presumably enzymologists will eventually adopt the same solution and replace the 'unit' with the nkat.

The curve defined by equation 2.7 is shown in *Figure 2.1*. It is a rectangular hyperbola through the origin, with asymptotes $s = -K_m$

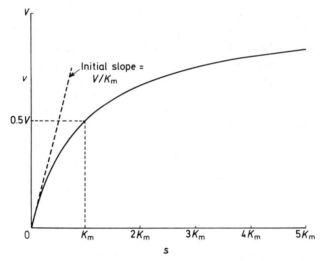

Figure 2.1 Plot of initial rate v against substrate concentration s for a reaction obeying the Michaelis–Menten equation. This plot should *not* be referred to as a 'Michaelis–Menten plot', as it was not used or advocated by Michaelis and Menten

and $v = V$. At very small values of s the denominator of equation 2.7 is dominated by K_m, i.e. s is negligible compared with K_m, and v is directly proportional to s:

$$v \simeq \frac{Vs}{K_m}$$

and the reaction is approximately first order in s. It is instructive to realize that V/K_m has this fundamental meaning as the first-order rate constant for the reaction $E + S \rightarrow E + P$ at low substrate concentrations, and that it should not be regarded solely as the result of dividing V by K_m. (The corresponding second-order rate constant, applicable when E is regarded as a reactant in this reaction, is k_{cat}/K_m.) When s is equal to K_m, equation 2.7 simplifies to $v = Vs/2s = 0.5V$, i.e. the rate is *half-maximal.* At very large values of s, the denominator of equation 2.7 is dominated by s, i.e. K_m is negligible in comparison with s, and the equation simplifies to

$$v \simeq V$$

i.e. the reaction is approximately zero order in s; under these conditions the enzyme is said to be *saturated.*

Unfortunately the plot of v against s is presented in a highly misleading way in many textbooks of biochemistry, and even in some specialist books on enzyme kinetics. As a result it is natural for students to gain a quite wrong impression of the shape of the curve, and to suppose that V can be estimated from such a plot of experimental observations by finding the point at which v 'reaches' its maximum value. The main fault lies in drawing a curve that flattens out too abruptly and then drawing an asymptote too close to the curve. In fact v never reaches V at finite values of s, and even when $s = 10K_m$ (a higher value than in many experiments) the value of v is still almost 10% less than V. This point may perhaps be grasped more clearly by examining a much greater proportion of the curve described by equation 2.7, as shown in *Figure 2.2*. This figure shows

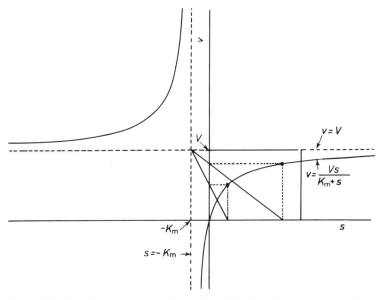

Figure 2.2   Plot of v against s according to the Michaelis–Menten equation. The part of the curve from $s = 0$ to $5K_m$ is the same as in *Figure 2.1*, but a much wider range of values is shown, including physically impossible values, to display the relationship of the curve to its asymptotes $s = -K_m$ and $v = V$

not only the usual range of s from 0 to a few times K_m, but a much wider range including physically impossible negative values. This explains the relationship of the curve to the usual two-limb hyperbolas found in textbooks of mathematics, and it also shows that when one estimates K_m and V from a set of observations one is in effect trying to locate the whole of an infinite curve, and the point of intersection of the asymptotes, from observations along a short arc. It is for this reason that estimation of K_m and V is not a trivial problem but one that requires considerable care. I shall return to it in Section 2.5, and again in Chapter 10.

It is tempting to assume that K_m can be taken as a measure of the true binding constant, K_s, in practice, i.e. to assume that k_{+2} is negligible in comparison with k_{-1}. In reality there is no justification for such an assumption unless supported by other evidence, and there are numerous examples of enzymes for which it is known to be untrue. Moreover there are many mechanisms with more than two steps that generate a rate equation of the same form as equation 2.7. For these the expression for K_m is complicated and does not necessarily simplify to K_s under plausible conditions. So in general K_m should be regarded as an empirical quantity that describes the dependence of v on s; it should not be taken as a measure of the thermodynamic stability of the enzyme–substrate complex.

2.4 Validity of the steady-state assumption

In the derivation of equation 2.7 it was necessary to introduce the assumption that a steady state would be reached in which $dx/dt = 0$. In fact, however, equation 2.3 is readily integrable and it is instructive to derive a rate equation without making the steady-state assumption, because this sheds some light on the validity of the assumption. Separating the two variables, x and t, we have

$$\int \frac{dx}{k_{+1} e_0 s - (k_{+1} s + k_{-1} + k_{+2})x} = \int dt$$

In spite of its more complicated appearance, the left-hand side is of the same simple form as several integrals we have encountered already (e.g. in Section 1.1), and may be integrated in the same way:

$$\frac{\ln [k_{+1} e_0 s - (k_{+1} s + k_{-1} + k_{+2})x]}{-(k_{+1} s + k_{-1} + k_{+2})} = t + \alpha$$

At the instant when the reaction starts, there can be no intermediate, i.e. $x = 0$ when $t = 0$, and so

$$\alpha = \frac{\ln (k_{+1} e_0 s)}{-(k_{+1} s + k_{-1} + k_{+2})}$$

giving

$$\ln \left[\frac{k_{+1} e_0 s - (k_{+1} s + k_{-1} + k_{+2})x}{k_{+1} e_0 s} \right] = -(k_{+1} s + k_{-1} + k_{+2})t$$

Taking exponentials of both sides, we have

$$1 - \frac{(k_{+1} s + k_{-1} + k_{+2})x}{k_{+1} e_0 s} = \exp [-(k_{+1} s + k_{-1} + k_{+2})t]$$

which may be rearranged to give the following expression for x:

$$x = \frac{k_{+1} e_0 s\{ 1 - \exp [-(k_{+1} s + k_{-1} + k_{+2})t]\}}{k_{+1} s + k_{-1} + k_{+2}}$$

The velocity is given by $k_{+2}x$, and so, substituting $V = k_{+2}e_0$ and $K_m = (k_{-1} + k_{+2})/k_{+1}$, we have

$$v = \frac{Vs\{1 - \exp[-(k_{+1}s + k_{-1} + k_{+2})t]\}}{K_m + s} \tag{2.8}$$

When t becomes very large the exponential term approaches $\exp(-\infty)$, i.e. zero, and equation 2.8 becomes identical to equation 2.7, i.e. the Michaelis–Menten equation. How large t must be for this to happen depends on the magnitude of $(k_{+1}s + k_{-1} + k_{+2})$: if we assume it to be of the order of $1000\ s^{-1}$ (a reasonable value in practice), then the exponential term is of the order of $\exp(-1000t)$, which is less than 0.01 for values of t greater than 5 ms.

   In the derivation of equation 2.8, s was treated as a constant, which is not strictly correct as s must change as the reaction proceeds. However, provided that s_0 is much larger than e_0, as is usually the case in steady-state experiments, the variation of s during the establishment of the steady state is trivial and can be neglected without significant inaccuracy. Laidler (1955) derived an equation similar to equation 2.8 as a special case of a much more general treatment in which he allowed for s to decrease from its initial value of s_0. He found that a steady state was achieved in which

$$v = \frac{V(s_0 - p)}{K_m + s_0 - p} \tag{2.9}$$

which is the same as equation 2.7 apart from the replacement of s with $s_0 - p$. It may seem contradictory to refer to a steady state in which v must decrease as p increases, but this decrease in v is extremely slow compared with the very rapid increase in v that occurs in the *transient phase*, i.e. the period before the steady state is established when equation 2.8 applies. The argument used by Briggs and Haldane is very little affected by replacing the assumption that $dx/dt = 0$ with an assumption that dx/dt is very small: equation 2.4 becomes a good approximation instead of an exact statement. As Wong (1975) has pointed out, what matters is not the absolute magnitude of dx/dt but its magnitude relative to $k_{+1}e_0s$.

2.5 Graphical representation of the Michaelis–Menten equation

If a series of initial velocities is measured at different substrate concentrations, it is desirable to present the results graphically, so that the values of the kinetic parameters can be estimated and the precision of the experiment assessed. The most obvious way of plotting equation 2.7 is to plot v against s, as in *Figure 2.1*. This is a most unsatisfactory plot in practice, however, for several reasons: it is difficult to draw a rectangular hyperbola accurately; it is difficult to locate the asymptotes correctly (because one is tempted to place them too close to the curve); it is difficult to perceive the relationship between a family of hyperbolas; and it is difficult to detect deviations from the expected curve if they occur. These disadvantages were recognized by Michaelis and Menten (1913), who instead plotted v against log s. Their plot has some advantages as well as its historical interest, but it is not now in common use and I shall not discuss it here.

Most workers since Lineweaver and Burk (1934) have preferred to rewrite the Michaelis–Menten equation in a way that permits the results to be plotted as points on a straight line. The way most commonly used is obtained from equation 2.7 by taking reciprocals of both sides:

$$\frac{1}{v} = \frac{1}{V} + \frac{K_m}{V_s} \tag{2.10}$$

This equation shows that a plot of $1/v$ against $1/s$ should be a straight line with slope K_m/V and intercept $1/V$ on the $1/v$ axis. This plot, which is commonly known as the *Lineweaver–Burk* or *double-reciprocal plot*, is illustrated in *Figure 2.3*. In spite of its great popularity, this plot cannot be recommended because it gives a grossly misleading impression of the experimental error: for small values of v, small errors in v lead to enormous errors in $1/v$; but for large values of v the same small errors in v lead to barely noticeable errors in $1/v$. This may be judged from the error bars shown in *Figure 2.3*, each of which is drawn for the same error range in v[†].

If we multiply both sides of equation 2.10 by s, we obtain the equation for a much better plot:

$$\frac{s}{v} = \frac{K_m}{V} + \frac{s}{V} \tag{2.11}$$

This shows that a plot of s/v against s should also be a straight line, with slope $1/V$ and intercepts K_m/V on the s/v axis and $-K_m$ on the s axis. This plot, which is sometimes referred to as a *Hanes plot*, is illustrated in *Figure 2.4*. Over a fair range of s values the errors in

[†]In principle, these difficulties can be overcome by the use of suitable weights, but this solution is not altogether satisfactory because it often leads to a 'best-fit' line that is perceived by the eye as fitting very poorly. Incidentally, Lineweaver and Burk should not be blamed for the misuse of their plot by later workers: they were well aware of the need for weights and the methods to be used for determining them (*see* Lineweaver and Burk, 1934, and, especially, Lineweaver, Burk and Deming, 1934). One is led to wonder how many of the hundreds of experimenters who cite Lineweaver and Burk each year as authorities for an unweighted method have actually read what they wrote.

s/v provide a faithful reflection of those in v, as may be judged from the error bars in *Figure 2.4*, and for this reason the plot of s/v against s should be preferred over the other straight-line plots for most purposes.

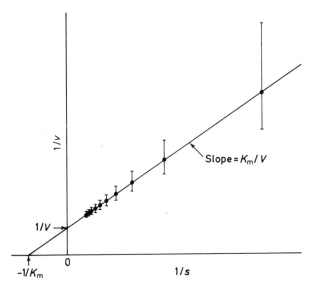

Figure 2.3 Plot of $1/v$ against $1/s$, with error bars of $\pm 0.05V$ in v (Lineweaver–Burk or double-reciprocal plot)

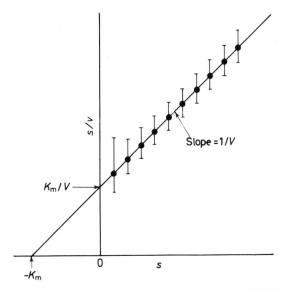

Figure 2.4 Plot of s/v against s, with error bars of $\pm 0.05V$ in v (sometimes called a Hanes plot)

Multiplying both sides of equation 2.10 by vV and rearranging, we obtain the equation for the third linear plot of the Michaelis–Menten equation:

$$v = V - \frac{K_m v}{s} \tag{2.12}$$

This shows that a plot of v against v/s should be a straight line with slope $-K_m$ and intercepts V on the v axis and V/K_m on the v/s axis. This plot, which is known as an *Eadie–Hofstee plot*, is illustrated in *Figure 2.5*. It gives fairly good results in practice, though the fact

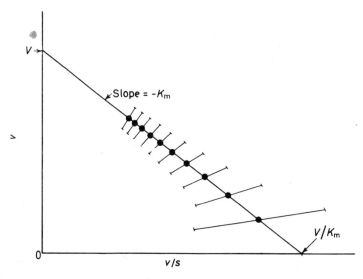

Figure 2.5   Plot of v against v/s, with error bars of $\pm 0.05V$ in v (Eadie–Hofstee plot)

that v appears in both co-ordinates means that errors in v cause deviations towards or away from the origin rather than parallel with the ordinate.

All three of these plots were first ascribed in print to Woolf (1932), but were not published by him. They became widely known and used as a result of the work of Lineweaver and Burk (1934), Eadie (1942) and Hofstee (1952), which is why some of them bear their names. Equation 2.11 was first published by Hanes (1932), but he did not present his results graphically.

A quite different way of plotting the Michaelis–Menten equation, known as the *direct linear plot*, has been described by Eisenthal and Cornish-Bowden (1974). The Michaelis–Menten equation may be rearranged in yet another way to show the dependence of V on K_m:

$$V = v + \frac{v}{s} K_m \tag{2.13}$$

(This equation can most simply be obtained by rearranging equation 2.12.) If V and K_m are treated as variables, and s and v as constants, this equation defines a straight line of slope v/s and intercepts v on the V axis and $-s$ on the K_m axis. Now it may seem perverse to treat V and K_m as variables and s and v as constants, but in fact it is more logical than it appears: once s and v have been measured in an experiment, they are constants, because any honest analysis of the results will leave them unchanged, but until we have decided on best-fit values of V and K_m we can try any values we like, and in that sense they are variables. For any pair of values s and v there is an infinite set of values of V and K_m that satisfy the values of s and v exactly. For any arbitrary value of K_m, equation 2.13 defines the corresponding value of V. Consequently, the straight line drawn according to equation 2.13 relates all pairs of K_m and V values that satisfy one observation exactly. If a second line is drawn for a second observation (with different values of s and v), it will relate all pairs of K_m and V values that satisfy the second observation exactly. However, the two lines will not define the same pair of K_m and V values except at the point of intersection. This point therefore defines the unique pair of K_m and V values that satisfies *both* observations exactly. If one lived in an ideal world in which there were no experimental error, then one could plot a series of such lines, each corresponding to a

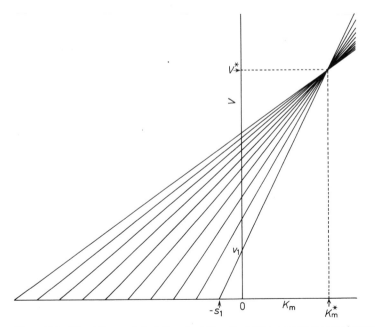

Figure 2.6 Direct linear plot of V against K_m. Each line represents one observation, and is drawn with intercepts $-s$ on the abscissa and v on the ordinate. The point of intersection gives the co-ordinates of the values of K_m and V that fit the data

single determination of v at a particular value of s, and they would all intersect at a common point, which would define the values of K_m and V that gave rise to the observations. Such an idealized plot is illustrated in *Figure 2.6*. In a real experiment the point of inter-section would be less well-defined than that shown in *Figure 2.6*, on account of experimental error, but it should normally be possible to define the best point as the point where the lines crowd closest together. I shall return to this matter in a more precise way in Chapter 10.

2.6 Reversible Michaelis–Menten mechanism

All reactions are reversible in principle, and many of the reactions of importance in biochemistry are also reversible in practice, in the sense that significant amounts of both substrates and products exist in the equilibrium mixture. It is evident, therefore, that the Michaelis–Menten mechanism, as given, is incomplete, and that allowance should be made for the reverse reaction

$$E + A \underset{k_{-1}}{\overset{k_{+1}}{\rightleftharpoons}} EA \underset{k_{-2}}{\overset{k_{+2}}{\rightleftharpoons}} E + P$$

$$e_0 - x\ a \qquad\qquad x \qquad\qquad\quad p$$

(When we are discussing mechanisms in which we are interested in more than one substrate, it is convenient not to use the symbol S for any particular substrate, but to reserve it for substrates in general. In these cases I shall use A, B, . . . for the substrates of the forward reaction and P, Q, . . . for the substrates of the reverse reaction.) The steady-state assumption is now expressed by

$$\mathrm{d}x/\mathrm{d}t \;=\; k_{+1}(e_0 - x)a + k_{-2}(e_0 - x)p - (k_{-1} + k_{+2})x \;=\; 0$$

Gathering together terms in x and rearranging, we obtain

$$x \;=\; \frac{k_{+1}e_0 a + k_{-2}e_0 p}{k_{-1} + k_{+2} + k_{+1}a + k_{-2}p}$$

Because this is a reversible reaction, to get the *net* rate of release of P we must subtract the rate at which P is consumed in the reaction $E + P \rightarrow EA$ from the rate at which it is released in the reaction $EA \rightarrow E + P$:

$$v \;=\; k_{+2}x - k_{-2}(e_0 - x)p$$

$$= \frac{k_{+2}(k_{+1}e_0 a + k_{-2}e_0 p)}{k_{-1} + k_{+2} + k_{+1}a + k_{-2}p} - k_{-2}e_0 p + \frac{k_{-2}(k_{+1}e_0 a + k_{-2}e_0 p)p}{k_{-1} + k_{+2} + k_{+1}a + k_{-2}p}$$

Cross-multiplication to express everything over the same denominator gives an apparently complicated numerator with eight terms. However, six of these cancel out and we are left with

$$v = \frac{k_{+1}k_{+2}e_0a - k_{-1}k_{-2}e_0p}{k_{-1} + k_{+2} + k_{+1}a + k_{-2}p} \tag{2.14}$$

The special case $p = 0$ gives the same equation as before, i.e. equation 2.6, except that a should be replaced with a_0, because the initial-rate condition is satisfied only at zero time. It is important to understand that this simplification is possible because p is zero, *not* because of any assumption about the magnitude of k_{-2}: the initial-rate condition applies when $p = 0$ because $k_{-2}p$ is zero if $p = 0$, regardless of the value of k_{-2}.

Equation 2.14 simplifies to a complementary special case for the initial rate of the reverse reaction if $a = 0$:

$$v = \frac{-k_{-1}k_{-2}e_0p_0}{k_{-1} + k_{+2} + k_{-2}p_0}$$

The negative sign in this equation arises because we have defined the rate as the rate of release of P, i.e. dp/dt; if we had defined it as da/dt the rate would have turned out to be positive. Apart from the sign, this equation is of the form of the Michaelis–Menten equation (equation 2.7), and we can define a maximum velocity and Michaelis constant for the reverse reaction:

$$V^r = k_{-1}e_0$$

$$K_m^P = (k_{-1} + k_{+2})/k_{-2}$$

which are analogous to the corresponding definitions for the forward reaction:

$$V^f = k_{+2}e_0$$

$$K_m^A = (k_{-1} + k_{+2})/k_{+1}$$

Using these four definitions, we can rewrite equation 2.14 as follows:

$$v = \frac{\dfrac{V^f a}{K_m^A} - \dfrac{V^r p}{K_m^P}}{1 + \dfrac{a}{K_m^A} + \dfrac{p}{K_m^P}} \tag{2.15}$$

This equation can be regarded as the general reversible form of the Michaelis–Menten equation. It has the advantage over equation 2.14

that it does not imply a particular mechanism and can be regarded as purely empirical: there are many mechanisms more complicated than the one at the beginning of this section that can be described by equation 2.15. The most important of these is the more realistic reversible mechanism in which the conversion of A into P is distinguished from the release of P from the enzyme:

$$E \; + \; A \underset{k_{-1}}{\overset{k_{+1}}{\rightleftharpoons}} EA \underset{k_{-2}}{\overset{k_{+2}}{\rightleftharpoons}} EP \underset{k_{-3}}{\overset{k_{+3}}{\rightleftharpoons}} E \; + \; P$$

$$e_0 - x - y \qquad\qquad x \qquad\qquad y \qquad\qquad p$$

In principle we can derive a rate equation for this mechanism by the same method as before. However, now there are two intermediates, and both dx/dt and dy/dt must be set to zero. Two simultaneous equations in x and y must be solved and the derivation is rather complicated. As I shall be describing a much more versatile method for deriving rate equations in Chapter 4, I shall simply state here that the three-step mechanism again leads to equation 2.15, but the definitions of the parameters are now

$$V^{\mathrm{f}} \; = \; \frac{k_{+2} k_{+3} e_0}{k_{-2} + k_{+2} + k_{+3}}$$

$$V^{\mathrm{r}} \; = \; \frac{k_{-1} k_{-2} e_0}{k_{-1} + k_{-2} + k_{+2}}$$

$$K_{\mathrm{m}}^{\mathrm{A}} \; = \; \frac{k_{-1} k_{-2} + k_{-1} k_{+3} + k_{+2} k_{+3}}{k_{+1} (k_{-2} + k_{+2} + k_{+3})}$$

$$K_{\mathrm{m}}^{\mathrm{P}} \; = \; \frac{k_{-1} k_{-2} + k_{-1} k_{+3} + k_{+2} k_{+3}}{(k_{-1} + k_{-2} + k_{+2}) k_{-3}}$$

In spite of their complex appearance, the expressions for $K_{\mathrm{m}}^{\mathrm{A}}$ and $K_{\mathrm{m}}^{\mathrm{P}}$ simplify to the true dissociation constants $K_{\mathrm{s}}^{\mathrm{A}}$ and $K_{\mathrm{s}}^{\mathrm{P}}$ of EA and EP, respectively, if the second step of the reaction is rate-limiting in either direction. For example, $K_{\mathrm{m}}^{\mathrm{A}} = k_{-1}/k_{+1} = K_{\mathrm{s}}^{\mathrm{A}}$ if k_{+2} is small compared with $(k_{-2} + k_{+3})$, $K_{\mathrm{m}}^{\mathrm{P}} = k_{+3}/k_{-3} = K_{\mathrm{s}}^{\mathrm{P}}$ if k_{-2} is small compared with $(k_{-1} + k_{+2})$; and both may be true simultaneously if $k_{+2} \ll k_{+3}$ and $k_{-2} \ll k_{-1}$, i.e. if the interconversion of EA and EP is rate-limiting in both directions. It is also possible for both Michaelis constants to be equilibrium constants without either of the binding steps needing to be at equilibrium: if $k_{-1} = k_{+3}$ (i.e. if A and P are released from their respective complexes with the same rate constant), then both expressions have common factors of $(k_{-2} + k_{+2} + k_{+3})$ in numerators and denominators, so

$K_m^A = k_{-1}/k_{+1} = K_s^A$ and $K_m^P = k_{+3}/k_{-3} = K_s^P$ (Cornish-Bowden, 1976).

When a reaction is at equilibrium, the net velocity must be zero and, consequently, if a_∞ and p_∞ are the equilibrium values of a and p, it follows from equation 2.15 that

$$\frac{V^f a_\infty}{K_m^A} - \frac{V^r p_\infty}{K_m^P} = 0$$

and so

$$\frac{V^f K_m^P}{V^r K_m^A} = \frac{p_\infty}{a_\infty} = K$$

where K is the equilibrium constant of the reaction. This is an important result, and is known as the *Haldane relationship* (Haldane, 1930). It is true for any mechanism that is described by equation 2.15, not merely for the simple two-step Michaelis–Menten mechanism. More complex rate equations, such as those that describe reactions of several substrates, lead to more complex Haldane relationships, but for all equations there is at least one relationship of this type between the kinetic parameters and the equilibrium constant.

2.7 Product inhibition

Product inhibition is simply a special case of inhibition, which I shall discuss in detail in Chapter 5, but because it follows very naturally from the previous section it is convenient to mention it briefly here. When equation 2.15 applies, the rate must decrease as product accumulates, even if the decrease in substrate concentration is negligible, because the negative term in the numerator becomes relatively more important as equilibrium is approached, and because the third term in the denominator increases. In any reaction, the negative term in the numerator can be significant only if the reaction is significantly reversible. Now, in many essentially irreversible reactions, such as the classic example of the invertase-catalysed hydrolysis of sucrose, product inhibition is significant. This indicates that product must be capable of binding to the free enzyme and is compatible with the simplest two-step mechanism only if the *first* step is irreversible and the second is not. This does not seem very likely, at least as a general phenomenon. On the other hand, the three-step mechanism predicts that product inhibition can occur in an irreversible reaction if it is the second step, i.e. the chemical transformation, that is irreversible. In such a case, the accumulation of product causes the enzyme to be sequestered as the EP complex. For an irreversible reaction, equation 2.15 then becomes

$$v = \frac{V^f a / K_m^A}{1 + (a/K_m^A) + (p/K_s^P)} = \frac{V^f a}{K_m^A (1 + p/K_s^P) + a} \qquad (2.16)$$

K_m^P can legitimately be written as K_s^P if the reaction is irreversible, because if k_{-2} approximates to zero it must be small compared with $(k_{-1} + k_{+2})$.

Of course, the effect of added product should be the same as that of accumulated product, so one could measure initial rates with different concentrations of added product. For each product concentration, the initial rate for different substrate concentrations would obey the Michaelis–Menten equation, but with *apparent* values of V and K_m, given by (compare equation 2.16 with equation 2.7) $V^{app} = V^f$ and $K_m^{app} = K_m^A (1 + p/K_s^P)$. Thus V^{app} has the same value V^f as for the uninhibited reaction, but K_m^{app} is larger than K_m^A and increases linearly with p. In practice, product inhibition is sometimes of this type (e.g. the inhibition of invertase by fructose), but sometimes it is not (e.g. the inhibition of invertase by its other product, glucose — Michaelis and Pechstein, 1914). Moreover, there are many compounds other than products that inhibit enzymes. It is clear, then, that a more complete theory is required to account for these facts, which is developed in later chapters.

2.8 Integrated Michaelis–Menten equation

As I discussed at the beginning of this chapter, the early workers in enzyme kinetics encountered many difficulties because they followed the reaction over an extended period of time, and then tried to explain their observations in terms of integrated rate equations similar to those commonly applied in chemical kinetics. These difficulties were largely resolved when Michaelis and Menten (1913) showed that the behaviour of enzymes could be studied much more simply by measuring initial rates, when the complicating effects of product accumulation and substrate depletion did not apply. An unfortunate by-product of this early history, however, has been that biochemists have been reluctant to use integrated rate equations, even when they have been appropriate. It is not always possible to carry out steady-state experiments in such a way that the progress curve (i.e. a plot of p against t) is essentially straight for an appreciable period, and in such cases estimation of the initial slope, and hence the initial rate, is subjective and liable to be biased. Much of this subjectivity and bias can be removed by using an integrated form of the rate equation, as I shall now describe.

If we write the Michaelis–Menten equation in its usual form as $v = dp/dt = Vs/(K_m + s)$ (cf. equation 2.7), it is an equation in three variables, p, t and s. As such it cannot be integrated directly, but one of the three variables can readily be removed by means of the stoicheiometric relationship $s + p = s_0$. Then we have

$$\frac{dp}{dt} = \frac{V(s_0 - p)}{K_m + s_0 - p}$$

which may be integrated by separating the two variables on to the two sides of the equation:

$$\int \frac{(K_m + s_0 - p)\, dp}{s_0 - p} = \int V\, dt$$

The left-hand side of this equation is not immediately recognizable as a simple integral, but, separating it into two terms, we have

$$\int \frac{(K_m + s_0)\, dp}{s_0 - p} - \int \frac{p\, dp}{s_0 - p} = \int V\, dt$$

of which the first is of the standard form that we have used several times before (e.g. in Section 1.1), and the second is of the standard form $\int x\, dx/(A + Bx) = (x/B) - (A/B^2) \ln (A + Bx)$, and so

$$-(K_m + s_0) \ln (s_0 - p) + p + s_0 \ln (s_0 - p) = Vt + \alpha$$

where α is a constant of integration that may be evaluated by means of the boundary condition $p = 0$ when $t = 0$, i.e. $\alpha = -K_m \ln s_0$. So, after substituting this value of α and rearranging, we have

$$Vt = p + K_m \ln [s_0/(s_0 - p)]$$

This may be regarded as the integrated form of the Michaelis–Menten equation, but for most purposes it is better to write it as

$$V^{app} t = s_0 - s + K_m^{app} \ln (s_0/s) \qquad (2.17)$$

where p has been written as $s_0 - s$ and V and K_m have been replaced by apparent values V^{app} and K_m^{app} respectively. The reason for these substitutions is that equation 2.17 applies much more generally than simply to the case for which we have derived it, for example to reactions subject to competitive product inhibition (equation 2.16), and in these cases V^{app} and K_m^{app} are not equal to V and K_m. Indeed, V^{app} and K_m^{app} can both be negative in some circumstances, whereas V and K_m are always positive. Nonetheless, V^{app} and K_m^{app} are capable of yielding highly accurate values of the initial rate v_0, by means of the equation

$$v_0 = V^{app} s_0/(K_m^{app} + s_0) \qquad (2.18)$$

even when they are themselves grossly poor estimates of V and K_m (Cornish-Bowden, 1975). Equation 2.18 follows from equation 2.17 by differentiation, regardless of the meanings of V^{app} and K_m^{app}.

Rearranging equation 2.17, we have

$$\frac{t}{\ln (s_0/s)} = \frac{1}{V^{app}} \left[\frac{s_0 - s}{\ln (s_0/s)} \right] + \frac{K_m^{app}}{V^{app}} \tag{2.19}$$

which shows that a plot of $t/\ln (s_0/s)$ against $(s_0 - s)/\ln (s_0/s)$ gives a straight line of slope $1/V^{app}$ and intercept K_m^{app}/V^{app} on the ordinate. V^{app} and K_m^{app} can readily be determined from such a plot, and v_0 can be calculated from them by means of equation 2.18. However, v_0 may also be found directly without evaluating V^{app} and K_m^{app} by a simple extrapolation of the line: rearrangement of equation 2.18 into the form of equation 2.11 shows that the point $(s_0, s_0/v_0)$ should lie on a straight line of slope $1/V^{app}$ and intercept K_m^{app}/V^{app} on the ordinate, i.e. the same straight line as that plotted from equation 2.19. Accordingly, if that line is extrapolated back to a point at which $(s_0 - s)/\ln (s_0/s) = s_0$, the value of the ordinate must be s_0/v_0. The whole procedure is illustrated in *Figure 2.7*. The extrapolated point can then be treated as a point on an ordinary plot of

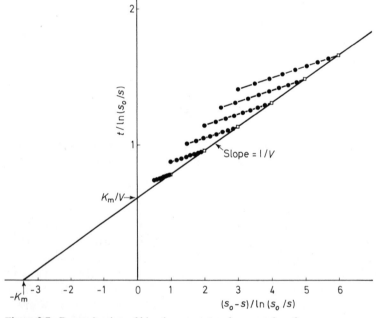

Figure 2.7 Determination of kinetic parameters from a series of progress curves at different values of the initial substrate concentration, s_0, by plotting $t/\ln (s_0/s)$ against $(s_0 - s)/\ln (s_0/s)$. For each value of s_0, the open square was obtained by extrapolating the line through the experimental points (filled circles) back to $(s_0 - s)/\ln (s_0/s) = s_0$, i.e. to 0% of reaction. These extrapolated points lie on a straight line of slope $1/V$ and intercepts $-K_m$ and K_m/V on the abscissa and ordinate respectively (cf. *Figure 2.4*)

s_0/v_0 against s_0 (*Figure 2.4*), and if several such points are found
from several progress curves with different values of s_0, K_m and V
may be found as described previously (Section 2.5).

This procedure, which originated with Jennings and Niemann
(1955), may seem an unnecessarily laborious way of generating an
ordinary plot of s_0/v_0 against s_0, but it provides more accurate values
of s_0/v_0 than are available by more direct methods. The extrapolation
required is very short, and it can be carried out more precisely and
less subjectively than estimating the tangent of a curve extrapolated
back to zero time.

Problems

2.1   For an enzyme obeying the Michaelis–Menten equation, calculate (a) the substrate concentration at which $v = 0.1V$, (b) the
substrate concentration at which $v = 0.9V$, and (c) the ratio
between the two.

2.2   At the time of Victor Henri, it was not considered unreasonable
that an enzyme might act merely by its presence, i.e. without
necessarily entering into combination with its substrate. Show
that the following mechanism, in which ES is formed but is not
on the pathway between S and P, leads to a rate equation of the
same form as the Michaelis–Menten equation:

ES

$\updownarrow$

E + S $\longrightarrow$ E + P

2.3   In designing an activity assay for an enzyme, it is desirable for
the measured initial rate to be insensitive to small errors in the
substrate concentration. How large must s/K_m be if a 10% error
in s is transmitted to v as an error of less than 1%? (Assume that
the Michaelis–Menten equation is obeyed.)

2.4   In an investigation of the enzyme fumarase from pig heart, the
kinetic parameters for the forward reaction were found to be
$K_m = 1.7$ mM, $V = 2.5$ mM min^{-1}, and for the reverse reaction
$K_m = 3.8$ mM, $V = 1.1$ mM min^{-1}. Estimate the equilibrium
constant for the reaction between fumarate and malate. In an
experiment on a sample of fumarase from a different source,
the kinetic parameters were reported to be $K_m = 1.6$ mM,
$V = 0.24$ mM min^{-1} for the forward reaction and $K_m = 1.2$ mM,
$V = 0.12$ mM min^{-1}. Comment on the plausibility of this
report.

2.5   From the following data, estimate the initial velocity v_0 at each initial substrate concentration s_0 from plots of product concentration p against time t (i.e. do *not* use the more elaborate method described in Section 2.8). Hence estimate K_m and V, assuming that the initial velocity is given by the Michaelis–Menten equation, by each of the methods illustrated in *Figures 2.1* and *2.3–2.6*. Finally, estimate K_m and V by the method described in Section 2.8. Account for any differences you observe between the results given by the different methods.

t (min)	p (mM)				
	$s_0 = 1$ mM	$s_0 = 2$ mM	$s_0 = 5$ mM	$s_0 = 10$ mM	$s_0 = 20$ mM
1	0.095	0.18	0.37	0.56	0.76
2	0.185	0.34	0.71	1.08	1.50
3	0.260	0.49	1.01	1.57	2.20
4	0.330	0.62	1.29	2.04	2.88
5	0.395	0.74	1.56	2.47	3.50
6	0.450	0.85	1.80	2.87	4.12
7	0.505	0.95	2.02	3.23	4.66
8	0.555	1.04	2.22	3.59	5.24
9	0.595	1.12	2.40	3.92	5.74
10	0.630	1.20	2.58	4.22	6.24

2.6   If a reaction is subject to product inhibition according to equation 2.16, the progress curve obeys an equation of the form

$$V^f t = (1 - K_m^A/K_s^P)(a_0 - a) + K_m^A (1 + a_0/K_s^P) \ln (a_0/a)$$

where a_0 is the value of a when $t = 0$ and the other symbols are defined as in equation 2.16. (a) Show that equation 2.16 is the differentiated form of this equation. (b) Compare the equation with equation 2.17 and write down expressions for V^{app} and K_m^{app} (defined as in equation 2.17). (c) Under what conditions will V^{app} and K_m^{app} be negative?

Chapter 3

Practical considerations

3.1 Purification of enzymes

Kafatos, Tartakoff and Law (1967) harvested cocoonase, a trypsin-like protease, as a dry, semicrystalline deposit of 80% pure enzyme, by removing it with forceps from the mouths of silk moths. Most of us, however, are less fortunate, because most enzymes occur naturally in a very impure state, and any serious attempt to characterize an enzyme must begin with a purification. Some aspects of enzymes — determination of the molecular weight and amino acid composition, for example — cannot be studied at all without a pure preparation; others, such as steady-state kinetic properties, can be studied with impure samples, but the conclusions that can be drawn are greatly limited if there is no information about the purity.

All enzymes are proteins, and there are marked chemical similarities between the most diverse of them. Furthermore, many enzymes are easily destroyed by treatments that the organic chemist would regard as very mild — heating to 100 °C, addition of organic solvents, etc. As a result, most of the techniques used for purifying simple organic compounds cannot be applied to enzymes. Instead it has been necessary to develop special techniques, most of which depend on the differential solubility of different enzymes in solutions of various compositions, or differential adsorption on solid materials. For pepsin, one of the first enzymes to be purified (Northrop, 1930), the effect of sulphuric acid on solubility was successfully exploited. Pepsin is an extracellular enzyme, however, and occurs in a relatively simple mixture; it is also much more stable than most other enzymes. More recently, therefore, milder treatments than addition of sulphuric acid have been developed. Most enzymes can be induced to precipitate, and under ideal conditions to crystallize, by addition of salts. The most widely used salt is ammonium sulphate, which causes most enzymes to precipitate without loss of catalytic activity. Precipitation by salt addition has the important advantage of concentrating the enzyme; most other purification techniques result in dilution and it is often advisable therefore to follow a diluting step in a purification scheme with precipitation by addition of ammonium sulphate.

Because of the chemical similarity between enzymes, it is not usually sufficient to rely on a single type of separation method; one must usually apply a series of different types of method. Enzymes differ from one another in their molecular size, their electric charge, their specificity for substrates and other small molecules, and their resistance to denaturing treatments such as heating; all of these properties can be exploited in separating them. I do not, however, intend to give a detailed description of experimental techniques here; my purpose is rather to consider the general principles of enzyme purification that apply to any enzyme.

3.2 Enzyme assays

In any purification it is essential to have an assay for catalytic activity, so as to be able to record the progress of the reaction as a function of time and hence determine the initial velocity. If it is unavoidable this can be a discontinuous assay, that is, one in which samples are removed at intervals from the reaction mixture and analysed to determine the extent of reaction. It is generally much more convenient, however, to devise a continuous assay, in which the progress of the reaction is monitored continuously with the aid of automatic recording apparatus. If the reaction causes a large change in absorbance at a conveniently accessible wavelength it can readily be followed in a recording spectrophotometer. For example, many reactions of biochemical interest involve the conversion of NAD^+ to NADH, and for these one can usually devise a spectrophotometric assay that exploits the large absorbance of NADH at 340 nm. Even if no such convenient spectroscopic change occurs in the reaction of interest it may well be possible to 'couple' it to a conveniently assayed reaction, as I shall discuss in Section 3.3.

Reactions for which no spectrophotometric assay is suitable may nonetheless often be followed continuously by taking advantage of the fact that many enzyme-catalysed reactions are accompanied by a release or consumption of protons. Such reactions may be followed in unbuffered solutions by means of a 'pH-stat', an instrument that adds base or acid automatically and maintains a constant pH. Because of the stoicheiometry of the reaction, a record of the amount of base or acid added provides a record of the progress of the reaction.

Ideally one must try to find conditions in which the progress curve is virtually straight for several minutes. In pedantic principle this is impossible, because regardless of the mechanism of the reaction one expects the rate to change — usually to decrease — as the substrates are consumed, the products accumulate and, often, the enzyme loses activity. A simple example of such slowing down is considered in Section 2.8, and some more complicated but more realistic ones in

Cornish-Bowden (1975). However, if it is possible to arrange the assay so that less than 1% of the complete reaction is followed, it may be true that the progress curve is indistinguishable from a straight line. This happy situation is not as common as one might suppose from reading the literature, because many experimenters are reluctant to recognize the inherent difficulty of drawing an accurate tangent to a curve, and prefer to persuade themselves that their progress curves are biphasic, with an initial 'linear' period followed by a tailing off. This nearly always causes the true initial velocity to be underestimated, for reasons that should be clear from *Figure 3.1*.

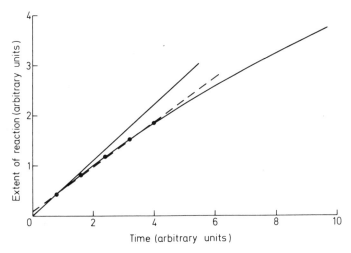

Figure 3.1 Bias in estimating an initial rate. The straight line through the origin is a true initial tangent, but the broken line, which has a slope about 20% smaller than that of the true tangent, is the line obtained by treating the five experimental points shown as if they occurred during the initial 'linear' part of the curve

To avoid the bias evident in *Figure 3.1*, the first essential is to be aware of the problem and to remember that one is trying to find the *initial* velocity, not the average velocity during the first few minutes of reaction. So one must try to draw a tangent to the curve extrapolated back to zero time, not a chord. For following the progress of a purification it is unlikely that any more refinement than this is necessary, as there is no need for highly precise initial velocities in this context. For subsequent kinetic study of the purified enzyme, however, one may well want better-defined initial velocities than one can hope to get by drawing initial tangents by hand. In this case a method based on an integrated rate equation as discussed in Section 2.8 is likely to be useful.

Provided that the rate of an enzyme-catalysed reaction is proportional to the total enzyme concentration, as one usually tries to ensure, one cannot alter the curvature of the progress curve by using more or less enzyme. One simply alters the scale of the time axis and any apparent change in curvature is an illusion; indeed this property is the basis of the test for enzyme inactivation described in Section 3.6. One can, however, improve the linearity of an assay by increasing the substrate concentration, so long as the products do not bind more tightly to the enzyme than the substrates (as often happens, for example, in reactions in which NAD^+ is converted into NADH). To illustrate this I shall consider the simplest possible case, that of a reaction that obeys the Michaelis–Menten equation and is not subject to product inhibition or retardation due to any cause apart from depletion of substrate. In this case the integrated form of the Michaelis–Menten equation (i.e. equation 2.17 with $V^{app} = V$ and $K_m^{app} = K_m$) describes the progress curve. If the initial substrate concentration s_0 is $5K_m$, the initial velocity v_0 is $0.83V$, and is largely independent of small errors in s_0. Moreover, if s_0 is doubled to $10K_m$, v_0 increases by only 9%, to $0.91V$. So it might seem that the assay would be insignificantly improved, and made considerably more expensive, by using the higher initial substrate concentration. But if non-linearity is a prime concern this conclusion is mistaken: a simple calculation (using equation 2.17) shows that if $s_0 = 5K_m$ the time taken for the rate to decrease by 1% is $0.34K_m/V$, but if $s_0 = 10K_m$ this time is more than trebled, to $1.11K_m/V$. In practice this calculation will usually be an oversimplification, because nearly all enzyme-catalysed reactions are subject to product inhibition, but the principle still applies qualitatively: increasing the initial substrate concentration usually extends the 'linear' period.

Another reason for using relatively high substrate concentrations in an enzyme assay is that the rate is then insensitive to small variations in substrate concentration, not only during the course of the reaction, as I have just discussed, but from one experiment to another. If one works with $s_0 = 0.1K_m$, for example, one must use precisely prepared solutions and work with great care, because a 10% error in s_0 will generate almost a 10% error in measured rate; when $s_0 = 10K_m$, on the other hand, much less precision is required because a 10% error in s_0 will generate an error of less than 1% in the measured rate.

3.3 Coupled assays

When it is not possible or convenient to follow a reaction directly in a spectrophotometer, it may nonetheless be possible to follow it indirectly by 'coupling' it to another reaction. Consider, for example, the hexokinase-catalysed transfer of a phosphate group from ATP to glucose:

glucose + ATP → glucose 6-phosphate + ADP

This important reaction is not accompanied by any convenient spectroscopic change, but it may nonetheless be followed spectrophotometrically by coupling it to the following reaction, which is catalysed by glucose 6-phosphate dehydrogenase:

glucose 6-phosphate + NAD^+ → 6-phosphogluconate + NADH

Provided that the activity of the coupling enzyme is high enough for the glucose 6-phosphate to be oxidized as fast as it is produced, the rate of NADH formation recorded in the spectrophotometer will correspond exactly to the rate of the reaction of interest.

The requirements for a satisfactory coupled assay may be expressed in simple but general terms by means of the scheme

$$A \xrightarrow{\;v_1\;} B \xrightarrow{\;v_2\;} C$$
$$a \qquad b \qquad c$$

in which the conversion of A into B at rate v_1 is the reaction of interest and the conversion of B into C is the coupling reaction, with a rate v_2 that can readily be measured. Plainly measurements of v_2 will provide accurate information about the initial value of v_1 only if a steady state in the concentration of B is reached before v_1 decreases perceptibly from its initial value. Most treatments of this system (e.g. McClure, 1969) assume that v_2 must have a first-order dependence on b, but this is both unrealistic and unnecessary, and may lead to the design of assays that are wasteful of materials. As the coupling reaction is usually enzyme-catalysed, it is more appropriate to suppose that the dependence of v_2 on b is given by the Michaelis–Menten equation:

$$v_2 = \frac{V_2 b}{K_2 + b} \tag{3.1}$$

where the symbols V_2 and K_2 are used to emphasize that they are the values of V and K_m for the second (coupling) enzyme. If v_1 is a constant (as is approximately the case during the period of interest, the early stages of reaction), the equation expressing the rate of change of b with time,

$$\frac{db}{dt} = v_1 - v_2 = v_1 - \frac{V_2 b}{K_2 + b} \tag{3.2}$$

can readily be integrated (for details, *see* Storer and Cornish-Bowden, 1974). It leads to the conclusion that the time t required for v_2 to reach any specified fraction of v_1 is given by an equation of the form

$$t = \phi K_2 / v_1 \tag{3.3}$$

in which ϕ is a dimensionless number that depends only on the ratios v_2/v_1 and v_1/V_2. Values of ϕ that are likely to be useful in designing coupled assays are tabulated in *Table 3.1*. The only parameter that can be adjusted by the experimenter is V_2, because K_2 is fixed by the choice of coupling enzyme and the reaction conditions, and it must be made large enough for t to be small in relation to the period of the assay.

TABLE 3.1   Time required for a coupled assay to reach a steady state

The table shows the value of ϕ to be inserted in equation 3.3 to give the time required for the rate v_2 measured in a coupled assay to reach 99% of the required rate v_1. For example, suppose v_1 is 0.1 mM min^{-1}, V_2, the maximum velocity of the coupling reaction, is 0.5 mM min^{-1}, and K_2, the Michaelis constant of the coupling enzyme under assay conditions, is 0.2 mM. Then the table gives $\phi = 1.31$; so the coupling system will take 2.62, i.e. 1.31 × 0.2/0.1, for the measured rate to reach 99% of the required rate. The table is abridged from Storer and Cornish-Bowden (1974).

v_1/V_2	ϕ	v_1/V_2	ϕ
0.0	0.00	0.5	6.86
0.1	0.54	0.6	11.7
0.2	1.31	0.7	21.4
0.3	2.42	0.8	45.5
0.4	4.12	0.9	141

The validity of this treatment can be checked by following the coupling reaction over a period of time and showing that the value of v_2 does increase in the expected way. An example of such a check is shown in *Figure 3.2*. In that experiment the value of V_2 was deliberately made rather smaller than would be appropriate for a satisfactory assay in order to make the period of acceleration clearly visible.

Even if the reaction of interest can be assayed directly, it is some-times advantageous to couple it to a second reaction. For example, if one of the products of the first reaction is a powerful inhibitor, or if a reversible reaction is being studied in the less favoured direction, so that equilibrium is reached after only a small percentage of sub-strate has reacted, it may be difficult to measure the initial velocity accurately. Problems of this kind can often be overcome by coupling the reaction to an irreversible reaction that removes the inhibitory product or displaces the equilibrium. In these cases much the same analysis as before applies, but it is advisable to define the steady state of the system rather more closely. The steady-state value of b in the scheme considered above is obtained by setting v_2 equal to v_i and solving equation 3.1 for b, which gives $b = K_2 v_1/(V_2 - v_1)$. It is then a simple matter to decide how large V_2 must be if the steady-state value of b is not to be large enough to cause problems.

Sometimes it is necessary to couple a reaction with two or more coupling enzymes. For example, the coupled assay mentioned above for hexokinase would not be satisfactory if one was trying to study inhibition of hexokinase by glucose 6-phosphate, because the

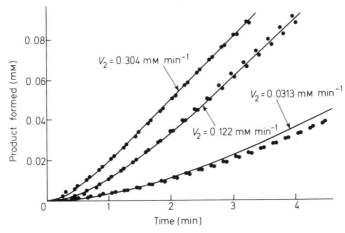

Figure 3.2 Data of Storer and Cornish-Bowden (1974) for the acceleration phase of the coupled assay for glucokinase, with glucose 6-phosphate dehydrogenase as coupling enzyme: the experimental points show (in duplicate) the concentrations of NADH, the product of the coupling reaction, at various times after the start of the reaction, and at three values of V_2 as indicated. The three curves are not fitted curves but theoretical curves calculated from the known values of V_2 according to the theory outlined in the text

coupling system would remove not only the glucose 6-phosphate released in the reaction but also any added by the experimenter. In this case one would need to couple the production of ADP to the oxidation of NADH, which normally requires two enzymes, pyruvate kinase and lactate dehydrogenase:

ADP + phospho-*enol*-pyruvate → pyruvate + ATP

pyruvate + NADH → lactate + NAD⁺

Rigorous kinetic analysis of systems with two or more coupling reactions is very difficult, but qualitatively they resemble the simple case we have considered: one must ensure that the activities of the coupling enzymes are high enough for the measured rate to reach 99% of the required rate within the time the required rate remains effectively constant. This can most easily be checked by experiment: if the enzyme concentrations are high enough there should be no effect on the measured rate if they are doubled.

3.4 Protein determination

Determination of protein concentration is only marginally within the scope of this book, but as it is essential in any purification I shall consider it briefly in this section. Layne (1957) may be consulted for a more detailed account of the most widely used methods.

The simplest and quickest method of estimating the concentration of protein in a sample is to measure the ultraviolet absorbance at 260 and 280 nm. This method, which was introduced by Warburg and Christian (1942), depends on the fact that nearly all proteins have an absorption maximum at about 280 nm (due mainly to tryptophan and tyrosine sidechains) and a minimum near 260 nm, whereas the reverse is true with nucleic acids. For example, for yeast enolase Warburg and Christian found the absorbance at 280 nm to be 1.75 times greater than at 260 nm, whereas for yeast nucleic acid the corresponding ratio was 0.49. As these are the main absorbing substances in the range 260–280 nm in typical cell extracts, it is possible to estimate the amount of protein from measurements at both wavelengths. Various detailed calculations have been proposed, but any attempt to make the method highly accurate is bound to be thwarted by the fact that yeast enolase and nucleic acid may not be typical of the protein and nucleic acid (and other ultraviolet absorbers) in the sample of interest. For most purposes therefore, the following equation, suggested by Layne (1957), is likely to be as accurate as it is worth while to expect:

$$\text{protein concentration (mg/ml)} = 1.55 A_{280} - 0.76 A_{260}$$

where A_{280} and A_{260} are the absorbances at 280 and 260 nm respectively.

The ultraviolet absorption method assumes that nucleic acid is the only contaminant in protein samples for which correction is required, and that the tryptophan and tyrosine content of any protein is about the same. Neither of these assumptions is exactly true, and whether they are acceptably accurate in any given case can only be determined by experiment. Nonetheless, the method has two important advantages over most others, which will ensure its continued use: it is very quick and convenient; and it does not involve destruction of the sample.

The most widely used alternative method is that proposed by Lowry et al. (1951), which combines the biuret reaction of protein with Cu^{2+} ions (so called because biuret is one of the few simple compounds that give the same reaction) with the reduction of phosphomolybdic-phosphotungstic acid by tyrosine and tryptophan sidechains. Both of these reactions lead to the development of a blue colour, which can be measured in a colorimeter. The advantage of combining the two methods is that they have complementary advantages and disadvantages: the biuret reaction is very specific for the peptide links in proteins and gives quantitatively similar results for each protein, but it is rather insensitive; the reduction of phospho-molybdic-phosphotungstic acid gives a much more intense colour, and is thus more sensitive, than the biuret method, but it is also more dependent on the particular proteins present, because of variations in tyrosine and tryptophan content. The complete method in the form proposed by Lowry *et al.* is more sensitive and more constant than the ultraviolet absorption method; but it is less convenient, it involves destruction of the (small) sample used for the estimation, and it gives a colour that is not strictly proportional to the concentration of protein in the sample.

Both of these methods are sometimes unsatisfactory, often because of interference by contaminants or substances added to stabilize the enzyme being studied. Consequently one should be aware of the many other methods that are available. Sober *et al.* (1965) list a dozen of these, and there are some others that are becoming more widespread. One is the micro-tannin turbidimetric method of Mejbaum-Katzenellenbogen and Dobryszycka (1959), which depends on the fact that under acid conditions tannic acid forms insoluble complexes with proteins. Like the biuret method, it depends on the peptide backbone of the protein and not on the sidechains, and it does not vary greatly in its sensitivity to different proteins. It is also cheap and convenient to carry out. More recently a method based on the dye-binding properties of proteins has been described (Bradford, 1976; Esen, 1978), which is more variable in its sensitivity (Pierce and Suelter, 1977; Van Kley and Hale, 1977), but still has considerable advantages in convenience and reproducibility over some of the better known methods.

In concluding this section I should emphasize that I have been concerned here with protein determination as a guide to the progress of a purification. For this purpose the variable sensitivity of most methods to different proteins is of little consequence. For characterizing a purified protein, however, much more rigorous methods are needed, as none of those mentioned in this section would be adequate. At this final stage, therefore, determinations of the molecular weight and amino acid composition should be made.

3.5 Presentation of results of a purification

In order for the experimenter and others to be able to judge the success of an enzyme purification, and to consider how it might be improved, it is essential to tabulate the results in a comprehensible way. Such a table must contain a brief indication of the chemical or

physical nature of each step, and a statement of the total amounts of
protein and enzyme activity after each step. This provides the minimum information required for judging the improvement afforded by
each step, but it is helpful and usual to add the following: the volume
of extract, the concentrations of protein and activity, the specific
activity (i.e. the amount of activity divided by the amount of protein), the yield, and the extent of purification. It is helpful to include
the volume in the table to enable others to judge the experimental
convenience of the procedure (it is of some interest to know in
advance whether one will be dealing with 5 ml or 50 l, for example!);
it is *essential* to measure the volume at each step, even if it is not to
be included in the table, because it is required for converting measured
concentrations into total amounts. The *specific activity* is a measure
of how much of the total protein is the enzyme required, and in a

TABLE 3.2   Purification of glucokinase from rat liver

The table shows two schemes for purifying glucokinase from rat liver. The first,
due to Parry and Walker (1966), involved only 'classical' techniques; results for
the livers of 10 rats are shown. The second scheme, due to Holroyde *et al.* (1976),
included affinity chromatography on glucosamine linked to Sepharose by a
6-aminohexanoyl 'spacer arm'. The results were obtained with 50 rats of a considerably larger size than those used by Parry and Walker, but are scaled down in
the table to give approximately the same starting weight of protein in both
schemes.

Stage	Protein (mg)	Volume (ml)	Total activity (μkat)	Specific activity (nkat/ mg)	Purification	Yield (%)
1 Liver supernatant	9 570	186	1.60	0.17	1	100
2 (NH$_4$)$_2$SO$_4$ fraction	2 700	37	†			
3 1st DEAE-Sephadex	170	103	0.83	4.9	29	52
4 2nd DEAE-Sephadex	31	18	0.72	23	140	45
5 DEAE-cellulose	12	63	0.53	44	260	33
6 Concentrated solution	3.0	2.0	0.24	80	480	15
7 Bio-Gel P225	1.8	22	0.24	130	780	15
8 Concentrated solution	1.1	2.2	0.16	145	870	10
1 Liver supernatant	9 440	156	0.87	0.092	1	100
2 DEAE-cellulose	45.3	18.9	0.91	20.1	220	104
3 Affinity chromato- graphy	1.73	7.22	0.73	420	4 500	83
4 DEAE-Sephadex	0.32	12.8	0.48	1 490	16 500	55
5 Concentrated solution	0.30	1.11	0.45	1 490	16 500	51
6 Sephadex G-200	0.14	8.89	0.36	2 560	28 000	42
7 Concentrated solution	0.14	0.56	0.35	2 510	28 000	40

†Activity could not be accurately measured at this stage because of uncertainty in correct-
ing for contamination by other enzymes.

successful purification it ought to increase at each step. The *yield* is the total enzyme activity after each step divided by the total activity of the initial extract: it will usually decrease at each step, but in a good purification the decrease should be small. The extent of *purification* is the specific activity after each step divided by the specific activity of the initial extract; it should increase at each step.

Table 3.2 shows two schemes for purifying glucokinase from rat liver. These illustrate not only the principles of constructing a purification table, which should require no further discussion, but also the enormous improvement in convenience, purity and yield that can be achieved by the use of affinity chromatography. Although some of the claims that have been made for this technique have been exaggerated – it is rarely possible to use it to the exclusion of the classical methods, for example – there can be little doubt that the effort required for designing an affinity column specific for a particular enzyme will usually be repaid in a much more effective and convenient purification scheme.

3.6 Detecting enzyme inactivation

Many enzymes are much more stable at high concentrations than at low, so it is not uncommon for an enzyme to lose activity rapidly when it is diluted from a stable stock solution to the much lower concentration used in the assay. If assay conditions can be devised that minimize inactivation the results are likely to be more reproducible than they would otherwise be, and in any case it is of interest to know whether the decrease in rate that occurs during the reaction is caused wholly or partly by loss of enzyme activity (rather than by substrate depletion or accumulation of products, for example). Fortunately Selwyn (1965) has described a simple test of this.

Selwyn has pointed out that as long as the rate dp/dt at all times during a reaction is proportional to the total enzyme concentration e_0 at the start, e_0 is a constant; then dp/dt can be expressed as the product of e_0 and some function of the instantaneous concentrations of the substrates, products, inhibitors and any other species that may be present. But because of the stoicheiometry of the reaction, these concentrations can in principle be calculated from p, the concentration of one product at any time. So the rate equation can be written in the simple form

$$\frac{dp}{dt} = e_0 \, f(p) \tag{3.4}$$

where f is a function that can in principle be derived from the rate equation. It is of no importance that f may be difficult to derive or that it may be a very complicated function of p, because its exact form is not required. It is sufficient to know that it is independent of e_0 and t, and so the integrated form of equation 3.4 must be

$$e_0 t = F(p)$$

where F is another function. The practical importance of this equation is that it shows that the value of $e_0 t$ after a specified amount of product has been formed is independent of e_0. Consequently, if progress curves are obtained with various values of e_0 but otherwise identical starting conditions, plots of $e_0 t$ against p for the various e_0 values should be superimposable. If they are not, the initial assumption that the rate throughout the reaction is proportional to the total enzyme concentration must be incorrect. *Figure 3.3* shows two examples of the use of this plot, one in which the results are as expected for a satisfactory assay, the other in which they are not.

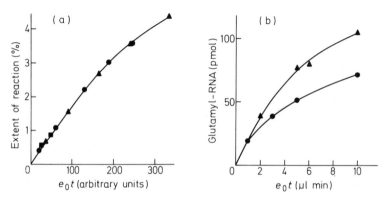

Figure 3.3   Selwyn's test of inactivation. In a reaction for which there is no appreciable inactivation during the time of observation, plots of the extent of reaction against $e_0 t$ should be superimposable, as in plot (a), which shows data of Michaelis and Davidsohn (1911) for invertase, at three different enzyme concentrations in the ratio 0.4 (■) : 1 (●) : 2 (▲). If the enzyme becomes inactivated during the reaction, or if the rate is not strictly proportional to e_0, the plots are not superimposable, as in plot (b), which shows data of Deutscher (1967) for glutamyl ribonucleic acid synthetase with 5.6 μg ml^{-1} (▲) and 2.8 μg ml^{-1} (●) enzyme

The simplest explanation of failure of Selwyn's test, as in *Figure 3.3b*, is that e_0 is not a constant because the enzyme becomes inactivated during the reaction. Selwyn lists several other possibilities, all of which indicate either that the assay is unsatisfactory or that it is complicated in some way that should be investigated before it is used routinely. For an example of this, *see* Problem 3.3 at the end of this chapter.

It is instructive to note that the principle embodied in Selwyn's test was widely known in the early years of enzymology: the data used for constructing *Figure 3.3a* were taken from Michaelis and Davidsohn (1911), and similar data are given by Hudson (1908); it is

clear moreover from the discussion given by Haldane (1930) that
similar tests were applied to many enzymes. As early as 1890,
O'Sullivan and Tompson commented that 'the time necessary to
reach any given percentage of inversion is in inverse proportion to
the amount of the inverting preparation present; that is to say, the
time is in inverse proportion to the inverting agent'. In spite of this,
the test was largely forgotten in modern times until Selwyn (1965)
adapted the treatment given by Michaelis and Davidsohn (1911) and
discussed the various reasons why the test might fail. One wonders
how many other useful techniques remain hidden in the early literature!

3.7 Experimental design: choice of substrate concentrations

A full account of the design of enzyme-kinetic experiments would
require a great deal of space, and in this section and the next I intend
to provide only a brief and simplified guide. In general, the conditions
that are optimal for assaying an enzyme, that is, determining the
amount of enzymic activity in a sample, are unlikely to be optimal
for determining its kinetic parameters. The reason for this is that in
an enzyme assay one tries to find conditions where the measured rate
depends *only* on the enzyme concentration, so that slight variations
in other conditions will have little effect; but in an investigation of
the kinetic properties of an enzyme one is concerned to know how it
responds to changes in conditions. It is essential in the latter case to
work over a range of substrate concentrations in which the rate varies
appreciably. In practice, for an enzyme that obeys the Michaelis–
Menten equation this means that s values should extend from below
K_m to well above K_m.

If one is confident that one is dealing with an enzyme that obeys
the Michaelis–Menten equation, one need only consider what range
of s values will define K_m and V precisely. It is easy to decide how to
define V precisely, by recalling that v approaches V as s becomes very
large (Section 2.3); obviously therefore it is desirable to include some
s values as large as expense and other considerations permit. Although
in principle the larger the largest s value is the better, in practice
there are two reasons why this may not be so. First, one's confidence
that the Michaelis–Menten equation is obeyed may be misplaced:
many enzymes show substrate inhibition at high s values, and as a
result the v values measured at very high s may not be those expected
from the K_m and V values that define the kinetics at low and moderate s. Second, even if the Michaelis–Menten equation is accurately
obeyed, the advantage of including s values greater than about $10K_m$
is very slight, and may well not be commensurate with the cost in
materials.

Just as the rate at high s is determined largely by V, so the rate at low s is determined largely by V/K_m (*see* Section 2.3). So for V/K_m to be precisely defined it is necessary for some observations to be made at s values less than K_m. It is not necessary to go to the lowest s values for which measurements are possible, however, because the need for v to be zero when s is zero provides a fixed point on the plot of v against s through which the curve must pass. As a result there is little advantage in using s values less than $0.5K_m$. To define K_m itself it is necessary to have accurate values of *both* V and V/K_m; thus one requires an s range from about $0.5K_m$ to $10K_m$ or as high as conveniently possible.

It cannot be overemphasized that the above remarks were prefaced with the condition that one must be confident that the Michaelis–Menten equation is obeyed, or alternatively that one does not care whether it is obeyed or not outside the range of the experiment. If one's interests are primarily physiological there is no reason why one should want to know about deviations from simple behaviour at grossly unphysiological concentrations; but if one is interested in enzyme mechanisms one should certainly explore as wide a range of conditions as possible, because deviations at the extremes of the experiment may well provide clues to the mechanism. Hill, Waight and Bardsley (1977) have argued that in reality there may be very few enzymes (if indeed there are any at all) that truly obey the Michaelis–Menten equation. They believe that excessively limited experimental designs, coupled with unwillingness to take note of deviations from expected behaviour, have led to an unwarranted belief that the Michaelis–Menten equation is almost universally obeyed. Nearly all of the standard and traditional examples of simple kinetics, they argue, prove on careful study to be more complex.

The Michaelis–Menten equation will undoubtedly remain useful as a first approximation in enzyme kinetics, even if it may sometimes need to be rejected after careful measurements, but it is always advisable to check for the most common deviations. Is the rate truly zero in the absence of substrate (and enzyme, for that matter)? If not, is the discrepancy small enough to be accounted for by instrumental drift or other experimental error? If there is a significant 'blank rate' in the absence of substrate or enzyme, can it be removed by careful purification? Does the rate approach zero at s values appreciably greater than zero? If so, it is worth while looking for evidence of co-operativity (Chapter 8). Is there any evidence of substrate inhibition, that is, decreasing v as s increases? Even if there is no decrease in v at high s values, failure to increase as much as expected from the Michaelis–Menten equation (*see Figure 2.1*) may indicate substrate inhibition.

3.8 Choice of pH, temperature and other conditions

Even if one does not intend to study the pH and temperature dependence of an enzyme-catalysed reaction, one must still give some
attention to the choice of pH and temperature. For many purposes
it will be appropriate to work under approximately physiological
conditions – pH 7.5, 37 °C, ionic strength 0.15 mol l^{-1} for most
mammalian enzymes, for example – but there may be good reasons
for deviating from these in a mechanistic study. Many enzymes
become denatured at an appreciable rate at 37 °C and may well be
much more stable at 25 °C (though there are exceptions, so this
should not be taken as a universal rule). It is also advisable to choose
a pH at which the reaction rate is insensitive to small changes in pH.
This is sometimes expressed in the form of a recommendation to
work at the pH 'optimum', but, as will become clear in Chapter 7,
this may well be meaningless advice unless K_m is independent of pH.
If K_m varies with pH, then even though the Michaelis–Menten equation may be obeyed, the maximum value of V/K_m will not occur at
the same pH as the maximum value of V; consequently the 'optimum'
pH will vary with the substrate concentration.

  In studies of reactions with more than one substrate, the experimental design must obviously be more complex than that required
for one-substrate reactions, but the principles are similar. Each substrate concentration should be varied over a wide enough range for
its effect on the rate to be manifest. If the Michaelis–Menten equation
is obeyed when any substrate concentration is varied under conditions
that are otherwise constant, the measured values of the Michaelis–
Menten parameters are *apparent* values, and are likely to change when
the other conditions are changed. To obtain the maximum information, therefore, one ought to use a range of substrate concentrations
chosen in relation to the appropriate apparent K_m, not the limiting
K_m as the other substrate(s) approach saturation, which may not be
relevant. I shall return to this topic in Chapter 6 after introducing the
basic equation for a two-substrate reaction. Similar considerations
apply in studies of inhibition and I shall discuss these in Chapter 5.

3.9 Use of replicate observations

At the end of a kinetic study one always finds that the best equation
one can determine fails to fit every observation exactly. The question
then arises as to whether the discrepancies are small enough to be
dismissed as experimental error, or whether they indicate the need
for a more complicated rate equation. To answer this question one
must have some idea of the magnitude of the random error in the

experiment. The clearest information about this can be obtained by including some repeated observations in the experiment. If the replicate observations agree with one another much better on average than they agree with the fitted line, there are grounds for rejecting the fitted line and perhaps introducing more terms into the equation. If, on the other hand, there is about as much scatter within each group of replicates as between the fitted line and the points, there can be no grounds for rejecting the equation until more precise observations become available.

The theory of this approach depends on the fact that it is only in a repeated experiment that one knows what the degree of agreement would be if there were no random error. Hence such an experiment measures only random error, or *pure error* as it may be called to distinguish it from the *lack of fit* caused by using an inadequate equation. The disagreement between an observation and a fitted line, on the other hand, may be caused either by error in the observation, or inadequacy of the theory, or most likely a combination of the two; it does not therefore measure pure error.

The use of repeated measurements is not without its pitfalls. To give a meaningful result the disagreement between replicates must be truly representative of the random error in the experiment as a whole. This will be true only if the repeated measurements are made just like any others, and not in any special way. This is perhaps best understood by examining the three examples shown in *Figure 3.4*. In *Figure 3.4a* the points are scattered within each group of replicates

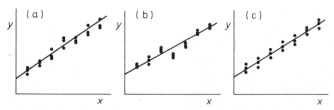

Figure 3.4 Use of repeated observations. When observations are properly repeated, the scatter of points about the fitted line should be irregular, as in (a). When the scatter is regular, as in (b) and (c), it suggests that the experiment has not been properly done, as discussed in the text

to much the same extent as all of the points are scattered about the line; this is what one expects when the repeated measurements have been made just like any others. In *Figure 3.4b* the scatter within each group of replicates is much less than the scatter about the line, even though the latter does appear to be random rather than systematic. There are various ways in which this kind of unsatisfactory result can arise: perhaps the commonest is to measure all of the observations within a group in succession, so that the average time between them is small compared with the average for the experiment. If this is done, any error caused by slow changes during the whole experiment — for example, instrumental drift, deterioration of stock solutions, increase in ambient temperature, fatigue of the experimenter — is not adequately manifest in the repeats.

Figure 3.4c shows the opposite problem. In this case the arrangement of each group of replicates is suspiciously regular, with a spread that is noticeably larger than the spread of points about the fitted line. This suggests that the repeats are overestimating the actual random error, perhaps because the figure actually represents three separate experiments done on three different days or with three different samples of enzyme.

The question of how many repeats there ought to be in a kinetic experiment is not one that can be answered dogmatically, though attempts to do so are sometimes made. The answer in any individual case must depend on how much work is required for each measurement, how long the enzyme and other stock solutions can be kept in an essentially constant state, how large the experimental error is, and how complicated the line to be fitted is. The first essential is to include as many *different* substrate (and inhibitor, etc.) concentrations as are needed to characterize the shape of the curve adequately. For a one-substrate enzyme that gave straightforward Michaelis–Menten kinetics, one might manage with as few as five substrate concentrations in the range $0.5K_m$ to $5K_m$; but with a two-substrate enzyme, again with straightforward kinetics, one might well require a minimum of 25 different combinations of concentrations; and for enzymes that show deviations from simple kinetics these numbers would certainly have to be increased. Only when the number of *different* concentrations to be used has been decided can one make an intelligent decision about the number of replicates required. Suppose one has decided that 25 different concentrations are necessary and that it is possible and convenient to measure 60 rates in the time available for the experiment, or the time during which deterioration of the enzyme is negligible. In such a case it would be appropriate to do ten sets of triplicates — spread over the whole experiment, not concentrated in one part of it — and the rest as duplicates. If on the other hand one could only manage 30 measurements one would have to decrease the number of repeats. To advocate a universal rule, that every measurement should be done in triplicate, for example, seems to me to be silly, not only because it oversimplifies the problem, but also because it may lead to experiments in which too few different sets of conditions are studied.

3.10 Treatment of ionic equilibria

For many reactions of biochemical interest, the substrate is not a well-characterized chemical compound with a directly measurable concentration, but an ion in equilibrium with other ions, some of which have their own interactions with the enzymes catalysing the reactions. Most notable of these ions is $MgATP^{2-}$, which is the true substrate of most of the enzymes that are loosely described as ATP-dependent. It is impossible to prepare a solution of pure $MgATP^{2-}$, because any solution that contains $MgATP^{2-}$ must also contain numerous other ions; for example, an equimolar mixture of ATP and $MgCl_2$ at pH 7 contains appreciable proportions of $MgATP^{2-}$, ATP^{4-}, $HATP^{3-}$, Mg^{2+} and Cl^-, as well as traces of $MgHATP^-$, Mg_2ATP and $MgCl^+$. Moreover, the proportions of these vary with the total ATP and $MgCl_2$ concentrations, the pH, the ionic strength and the concentrations of other species (such as buffer components) that may be present.

Obviously, if one is studying the effect of $MgATP^{2-}$, for example, on an enzyme, one requires some assurance that the effects attributed to $MgATP^{2-}$ are indeed due to that ion and not to variations in the Mg^{2+} and ATP^{4-} concentrations that accompany variations in the $MgATP^{2-}$ concentration. It is necessary, therefore, to have some method of calculating the composition of a mixture of ions, and it is desirable to have some way of varying the concentration of one ion without concomitant large variations in other concentrations.

The stability constants of many of the ions of biochemical interest have been measured. Thus it is a simple matter to calculate the concentration of any complex if the concentrations of the free components are known. Unfortunately, however, one usually encounters the problem in the converse form: given the total concentrations of the components of a mixture how can one calculate the free concentrations? Or, to take a specific example, given the total ATP and $MgCl_2$ concentrations, the pH and all relevant equilibrium constants, how can one calculate the concentration of $MgATP^{2-}$? A simple and effective approach is to proceed as follows:

(1)   Assume initially that no complexes exist and that all ionic components are fully dissociated. For example, assume that 1 mM ATP + 2 mM $MgCl_2$ + 100 mM KCl contains 1 mM ATP^{4-} 2 mM Mg^{2+}, 100 mM K^+ and 104 mM Cl^-.

(2)   Use these free concentrations and the association constants to calculate the concentrations of all complexes that contain Mg^{2+}. When added up, these will give a total Mg^{2+} concentration that exceeds (probably by a very large amount in the first stage of the calculation) the true total Mg^{2+} concentration.

(3)   Correct the free Mg^{2+} concentration by multiplying it by the true total Mg^{2+} concentration divided by the calculated total Mg^{2+} concentration.

(4) Repeat for each component in turn, i.e. for ATP^{4-}, Cl^- and K^+ in this example. In principle, H^+ may be treated in the same way, but in usual experimental practice the free H^+ concentration is controlled and measured directly and so the free H^+ concentration should not be corrected during the calculation but maintained throughout at its correct value.

(5) Repeat the whole cycle, steps (2)–(4), until the results are self-consistent, that is, until the concentrations do not change from cycle to cycle.

This procedure is slightly modified from one described by Perrin (1965) and Perrin and Sayce (1967). Although the number of cycles required for self-consistency is likely to be too large for convenient calculation by hand, the method is simple to express as a computer program (Storer and Cornish-Bowden, 1976a), and is then easy and efficient to apply to any of the problems likely to be encountered in enzyme kinetics.

Experience in using a program of this sort has led to a simple experimental design for varying the concentration of $MgATP^{2-}$ while keeping variations in the concentrations of other ions under control. Two designs are in common use, one of which gives good results, and the other of which leads to very poor results. The 'good' design is to keep the total $MgCl_2$ concentration in a constant excess over the total ATP concentration. The best results are obtained with an excess of about 5 mM $MgCl_2$, but if the enzyme is inhibited by free Mg^{2+}, or if there are other reasons for wanting to minimize the concentration of free Mg^{2+}, the excess can be lowered to 1 mM with only small losses of efficiency. If the excess is greater than 10 mM there may be complications due to the formation of Mg_2ATP in significant concentrations. With this design the ATP concentration may be varied over a wide range (1 μM to 0.1 M at least) with a high and almost constant proportion of the ATP existing as $MgATP^{2-}$ and a nearly constant concentration of free Mg^{2+}. Thus effects due to variation in the $MgATP^{2-}$ concentration may be clearly separated from effects due to variation in the free Mg^{2+} concentration.

The 'bad' design, which is fortunately encountered fairly infrequently in the literature, is to vary the total concentrations of ATP and $MgCl_2$ in constant ratio. Whether this ratio is 1 : 1 or any other, this design leads to wild variations in the proportion of ATP existing in any particular form, and cannot be recommended. Somewhat less objectionable, though still not to be recommended, is to keep the total $MgCl_2$ concentration constant at a value that exceeds the highest ATP concentration by about 2–5 mM. Although this design does ensure that ATP exists largely as $MgATP^{2-}$, it can produce undesirably large variations in the concentrations of free Mg^{2+} and of Mg_2 ATP.

Although the conclusions outlined in the preceding paragraphs depend to some extent on the numerical values of the equilibrium constants for complexes of Mg^{2+}, ATP^{4-} and H^+, the principles apply generally. As a rough guide, a component A of a binary complex AB exists largely in complexed form if B is maintained in excess over A by an amount about 100 times the dissociation constant of AB.

In this discussion I have simplified the problem by ignoring the fact that ionic equilibrium constants strictly define ratios of *activities* rather than concentrations. In practice, therefore, if one wants to avoid the complication of dealing with activity coefficients (in company with the overwhelming majority of biochemists), one must work at a constant ionic strength. A value of about 0.15 mol l^{-1} is appropriate, both because it is close to the ionic strength of many living cells, and because many of the equilibria of biochemical interest are insensitive to ionic strength near this value.

Problems

3.1   Brain hexokinase is strongly inhibited by glucose 6-phosphate at concentrations above 0.1 mM. What must the maximum velocity V_2 of glucose 6-phosphate dehydrogenase ($K_m = 0.11$ mM for glucose 6-phosphate) be if it is required as coupling enzyme in an assay for brain hexokinase in which rates v_1 not exceeding 0.1 mM min^{-1} are to be measured and the concentration of glucose 6-phosphate is never to exceed 0.1 mM?

3.2   Although steps in enzyme purification usually involve some loss of activity, one sometimes obtains yields of greater than 100% in single steps, especially in the early stages of purification. Suggest reasons why yields greater than 100% may occur.

3.3   The following data refer to two assays of the same enzyme, with identical reaction mixtures except that twice as much enzyme was added in (b) as in (a). Suggest a cause for the observed behaviour.

Time (min)	Concentration of product (μM)	
	(a)	(b)
0	0.0	0.0
2	10.5	4.3
4	18.0	8.3
6	23.7	11.7
8	27.9	14.5
10	31.3	16.8
12	34.0	19.0

3.4 The experimental design of maintaining a total $MgCl_2$ concentration 5 mM in excess of the total ATP concentration ensures that effects due to $MgATP^{2-}$ and Mg^{2+} can be clearly separated, because it allows the $MgATP^{2-}$ concentration to be varied with very little concomitant variation in free Mg^{2+} concentration. But it does not permit unequivocal distinction between effects of $MgATP^{2-}$ and of ATP^{4-}, because it causes their concentrations to be maintained in almost constant ratio. Suggest a design that would allow the $MgATP^{2-}$ concentration to be varied with very little variation in the ATP^{4-} concentration.

Chapter 4

How to derive steady-state rate equations

4.1 King–Altman method

In principle, the steady-state rate equation for any enzyme mechanism can be derived in the same way as that for the simple Michaelis–Menten mechanism: we write down expressions for the rates of change of concentrations of all of the intermediates, set them all equal to zero and solve the simultaneous equations that result. In practice, this method is extremely laborious and liable to error for all but the simplest mechanisms. Fortunately, King and Altman (1956) have described a schematic method that is simple to apply to any mechanism that consists of a series of reactions between different forms of one enzyme. It is not applicable to non-enzymic reactions, to mixtures of enzymes, or to reactions that contain non-enzymic steps. Nonetheless, it is applicable to most of the mechanisms met in enzyme catalysis and is very useful in practice. It is described and discussed in this chapter.

The method of King and Altman is most easily described by reference to an example, and I shall take as an example one of the most important two-substrate mechanisms:

$$E + A \underset{k_{-1}}{\overset{k_{+1}}{\rightleftharpoons}} EA$$

$$EA + B \underset{k_{-2}}{\overset{k_{+2}}{\rightleftharpoons}} EAB$$

$$EAB \rightleftharpoons EPQ$$

$$EPQ \underset{k_{-3}}{\overset{k_{+3}}{\rightleftharpoons}} EQ + P$$

$$EQ \underset{k_{-4}}{\overset{k_{+4}}{\rightleftharpoons}} E + Q$$

No rate constants are shown for the third reaction, because steady-state measurements provide no information about isomerizations between intermediates that react only in first-order reactions. For analytical purposes, therefore, we must treat EAB and EPQ as a single species, even though it may be mechanistically more meaningful to regard them as distinct.

The first step in the King–Altman method is to represent the mechanism by a scheme that shows all of the enzyme species and the reactions between them:

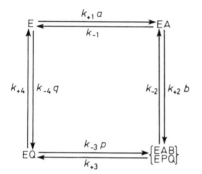

All of the reactions must be treated as first-order reactions. This means that second-order reactions, such as the reaction E + A → EA, must be given pseudo-first-order rate constants; for example, the second-order rate constant k_{+1} is replaced by the pseudo-first-order rate constant $k_{+1}a$ by including the concentration of A.

Next, a *master pattern* is drawn representing the skeleton of the scheme, in this case a square:

It is then necessary to find every pattern that (a) consists only of lines from the master pattern, (b) connects every enzyme species and (c) contains no closed loops. Each pattern will contain one line fewer than the number of enzyme species, and in this example there are four such patterns:

In this example the application of the rules is fairly obvious, but in a more complex mechanism it might not be and to avoid any misunderstanding it may be helpful to show three examples of improper patterns, each of which satisfies two but violates a third of the rules above:

For each enzyme species, arrowheads are then imagined on the lines of the patterns, in such a way that each pattern leads to the species considered, regardless of starting point. Thus for E, the four legitimate patterns would be imagined as follows:

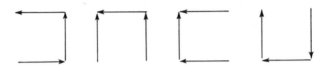

Then a sum of products of rate constants is written down for each enzyme species, such that each product contains the rate constants corresponding to the arrows that have been imagined. So, from the patterns leading to E, the sum of products is $(k_{-1}k_{-2}k_{-3}p + k_{-1}k_{-2}k_{+4} + k_{-1}k_{+3}k_{+4} + k_{+2}k_{+3}k_{+4}b)$. This sum is then the numerator of an expression that represents the fraction of the total enzyme concentration e_0 present as the species in question. So, for all four species, we have

$$[E]/e_0 = (k_{-1}k_{-2}k_{-3}p + k_{-1}k_{-2}k_{+4} + k_{-1}k_{+3}k_{+4} + k_{+2}k_{+3}k_{+4}b)/\Sigma$$

$$[EA]/e_0 =$$

$$(k_{+1}k_{-2}k_{-3}ap + k_{+1}k_{-2}k_{+4}a + k_{+1}k_{+3}k_{+4}a + k_{-2}k_{-3}k_{-4}pq)/\Sigma$$

$$[EAB + EPQ]/e_0 =$$

$$(k_{+1}k_{+2}k_{-3}abp + k_{+1}k_{+2}k_{+4}ab + k_{-1}k_{-3}k_{-4}pq + k_{+2}k_{-3}k_{-4}bpq)/\Sigma$$

$$[EQ]/e_0 =$$

$$(k_{+1}k_{+2}k_{+3}ab + k_{-1}k_{-2}k_{-4}q + k_{-1}k_{+3}k_{-4}q + k_{+2}k_{+3}k_{-4}bq)/\Sigma$$

The denominator Σ is the same for each species and is the sum of all four numerators, i.e. the sum of all 16 products obtained from the patterns.

   The rate of the reaction is then the sum of the rates of the steps that generate one particular product, minus the sum of the rates of the steps that consume the same product. In this example, there is one step only that generates P, $(EAB + EPQ) \rightarrow EQ + P$, and one step only that consumes P, $EQ + P \rightarrow (EAB + EPQ)$, so we have

$$v = dp/dt = k_{+3}[EAB + EPQ] - k_{-3}[EQ]p$$

$$= e_0(k_{+1}k_{+2}k_{-3}k_{+3}abp + k_{+1}k_{+2}k_{+3}k_{+4}ab$$

$$+ k_{-1}k_{-3}k_{+3}k_{-4}pq + k_{+2}k_{-3}k_{+3}k_{-4}bpq$$

$$- k_{+1}k_{+2}k_{-3}k_{+3}abp - k_{-1}k_{-2}k_{-3}k_{-4}pq$$

$$- k_{-1}k_{-3}k_{+3}k_{-4}pq - k_{+2}k_{-3}k_{+3}k_{-4}bpq)/\Sigma$$

$$= (k_{+1}k_{+2}k_{+3}k_{+4}e_0ab - k_{-1}k_{-2}k_{-3}k_{-4}e_0pq)/\Sigma$$

Notice that, although the method of King and Altman avoids generating any denominator terms that must subsequently be cancelled by subtraction, and thus avoids most of the wasted labour of the simpler approach, it does not succeed in avoiding cancellation in the numerator of the rate equation. In the present example, eight terms were derived for the numerator, of which six were then cancelled by subtraction. It is striking that the two numerator terms that remain have a rather tidy appearance compared with the six that disappeared: the positive term in the numerator consists of a product of the total enzyme concentration, all substrate concentrations for the forward reaction and all four rate constants for a complete cycle in the forward direction; the negative term in the numerator consists of a product of the total enzyme concentration, all substrate concentrations for the reverse reaction and all four rate constants for a complete cycle in the reverse direction. This result may be generalized into a rule for generating the numerators of rate equations without cancellation, as Wong and Hanes (1962) and Wong (1975) have described.

For most purposes it is more important to know the form of the steady-state rate equation than to know its detailed expression in terms of rate constants. For this reason it is often convenient to express a derived rate equation in *coefficient form*, which permits a straightforward prediction of the experimental properties of a given mechanism. For the example we have been examining, the coefficient form of the rate equation is

$$v = \frac{e_0(c_1ab - c_2pq)}{c_3 + c_4a + c_5b + c_6p + c_7q + c_8ab + c_9ap + c_{10}bq + c_{11}pq + c_{12}abp + c_{13}bpq}$$

where the coefficients c_1 to c_{13} have the following values:

$$c_1 = k_{+1}k_{+2}k_{+3}k_{+4}; c_2 = k_{-1}k_{-2}k_{-3}k_{-4}; c_3 = k_{-1}(k_{-2} + k_{+3})k_{+4};$$

$$c_4 = k_{+1}(k_{-2} + k_{+3})k_{+4}; c_5 = k_{+2}k_{+3}k_{+4}; c_6 = k_{-1}k_{-2}k_{-3};$$
$$c_7 = k_{-1}(k_{-2} + k_{+3})k_{-4}; c_8 = k_{+1}k_{+2}(k_{+3} + k_{+4}); c_9 = k_{+1}k_{-2}k_{-3};$$
$$c_{10} = k_{+2}k_{+3}k_{-4}; c_{11} = (k_{-1} + k_{-2})k_{-3}k_{-4}; c_{12} = k_{+1}k_{+2}k_{-3};$$
$$c_{13} = k_{+2}k_{-3}k_{-4}.$$

4.2 Modifications to the King–Altman method

The method of King and Altman as described is convenient and easy
to apply to any of the simpler enzyme mechanisms. However, complex mechanisms often require very large numbers of patterns to be
found. The derivation is then very laborious, and liable to errors on
account of patterns being overlooked or incorrect terms being written
down. Although it is possible in principle to calculate the total number of patterns (*see* King and Altman, 1956; Chou *et al.*, 1979), it is
very tedious unless the mechanism is very simple, because corrections
must be applied for all cycles within the mechanism. In any case,
knowing the number of patterns to be found may not be very helpful
in finding them, and does not reduce the labour involved in writing
down the terms. In general, for complex mechanisms, it is better to
search for ways of simplifying the procedure. Volkenstein and Goldstein (1966) have given a number of rules for doing this, of which the
simplest are the following:

(1) If there are two or more steps interconverting the same pair
of enzyme species, these steps can be condensed into one by
adding the rate constants for the parallel reactions. For example, the Michaelis–Menten mechanism is represented in the
King–Altman method as follows:

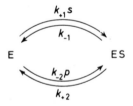

which gives two patterns, ⌢ and ⌣ . Because the
two reactions connect the same pair of enzyme species, they
can be added, to give

$$E \xrightleftharpoons[k_{-1} + k_{+2}]{k_{+1}s + k_{-2}p} ES$$

This scheme is itself the only pattern, and so the expressions
for [E] and [ES] can be written down directly:

$$[E]/e_0 = (k_{-1} + k_{+2})/(k_{-1} + k_{+2} + k_{+1}s + k_{-2}p)$$
$$[ES]/e_0 = (k_{+1}s + k_{-2}p)/(k_{-1} + k_{+2} + k_{+1}s + k_{-2}p)$$

In more complex cases, the simplification afforded by this technique is very great: an example discussed by King and Altman was the general modifier mechanism of Botts and Morales (1953):

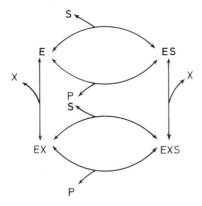

As shown, this master pattern requires twelve patterns, but if the parallel paths between E and ES and between EX and EXS are added, the master pattern becomes a square, which requires only four patterns.

(2) If the mechanism contains different enzyme species that have identical properties, the procedure is greatly simplified by treating them as single species. For example, if an enzyme contained two *identical* active sites, the mechanism might be represented as

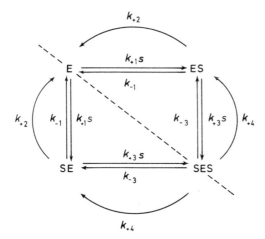

which requires 32 patterns. However, as ES and SE are identical, the pattern is symmetrical about the broken line, and can be represented much more simply as

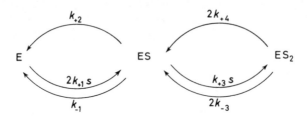

which can be further simplified by rule 1 to

$$E \underset{k_{-1} + k_{+2}}{\overset{2k_{+1} s}{\rightleftharpoons}} ES \underset{2(k_{-3} + k_{+4})}{\overset{k_{+3} s}{\rightleftharpoons}} ES_2$$

Thus a scheme of 32 patterns has been simplified to one of only a single pattern, and the expressions for the three species can be written down immediately:

$$[E]/e_0 = 2(k_{-1} + k_{+2})(k_{-3} + k_{+4})/[2(k_{-1} + k_{+2})(k_{-3} + k_{+4}) + 4k_{+1}(k_{-3} + k_{+4})s + 2k_{+1}k_{+3}s^2]$$

and so on.

Whenever advantage is taken of the symmetry of the master pattern in this way in order to condense it into a simpler scheme, statistical factors appear. In this example, the reaction E → ES can occur in two ways, so the total rate is the sum of the two rates, giving a rate constant $2k_{+1} s$ that is double the rate constant for either of the two paths. The reverse reaction, on the other hand, can occur in only one way, with a statistical factor of 1, and the rate constant remains k_{-1}.

(3)   If the master pattern consists of two or more distinct parts touching at single enzyme forms, it is convenient to treat the different parts separately. A simple example of this is provided by the case of *competitive substrates*, in which a single enzyme simultaneously catalyses two separate reactions with different substrates:

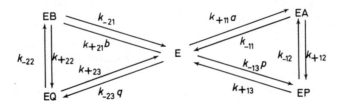

In this case, the expression for each enzyme form is the product of the appropriate sums for the left and right halves of the master pattern:

$$[E]/e_0 = (k_{+22}k_{+23} + k_{-21}k_{-22} + k_{-21}k_{+23})$$
$$\times (k_{+12}k_{+13} + k_{-11}k_{-12} + k_{-11}k_{+13})/\Sigma$$

$$[EA]/e_0 = (k_{+22}k_{+23} + k_{-21}k_{-22} + k_{-21}k_{+23})$$
$$\times (k_{-12}k_{-13}p + k_{+11}k_{-12}a + k_{+11}k_{+13}a)/\Sigma$$

$$[EP]/e_0 = (k_{+22}k_{+23} + k_{-21}k_{-22} + k_{-21}k_{+23})$$
$$\times (k_{+12}k_{-13}p + k_{+11}k_{+12}a + k_{-11}k_{-13}p)/\Sigma$$

$$[EB]/e_0 = (k_{-22}k_{-23}q + k_{+21}k_{-22}b + k_{+21}k_{+23}b)$$
$$\times (k_{+12}k_{+13} + k_{-11}k_{-12} + k_{-11}k_{+13})/\Sigma$$

$$[EQ]/e_0 = (k_{+22}k_{-23}q + k_{+21}k_{+22}b + k_{-21}k_{-23}q)$$
$$\times (k_{+12}k_{+13} + k_{-11}k_{-12} + k_{-11}k_{+13})/\Sigma$$

The expression for $[E]/e_0$ contains two sums, corresponding to the patterns that lead to E in the two halves of the mechanism. Of these two sums, the one on the left reappears in the expressions for EA and EP, whereas the one on the right reappears in the expressions for EB and EQ. The right-hand sum in the expression for EA corresponds to the patterns that lead to EA in the right half of the mechanism; and the expressions for EP, EB and EQ are composed similarly.

Volkenstein and Goldstein (1966) also described a fourth modification to the method of King and Altman, which provides one of the few satisfactory approaches to the analysis of complex mechanisms with branched pathways. However, it is much more difficult to understand and apply than the others, and it is not profitable to study it until one has had extensive practice with the method in its ordinary form. I shall therefore omit it from this treatment.

4.3 Cha's method for reactions containing steps at equilibrium

Some mechanisms are important enough to be worth analysing in detail, but so complex that even with the aid of the methods described above they give rise to unmanageably complicated rate equations. In such cases, some simplifying assumptions are unavoidable and great simplifications often result if one assumes that some steps, such as protonation steps, are maintained at equilibrium at all times. Such assumptions may, of course, turn out to be false after further investigation, but they are useful as a first approximation.

Cha (1968) has described a method for analysing mechanisms that contain some steps at equilibrium, which is much simpler than the full King–Altman analysis as each group of species at equilibrium can be treated as a single species. As an example, let us consider the general modifier mechanism of Botts and Morales (1953), with the assumption that binding of modifier X to both free enzyme and enzyme–substrate complex is maintained at equilibrium:

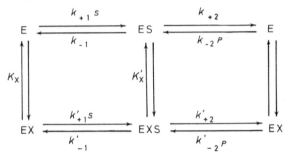

The rate of the reaction $E \rightarrow ES$ is $k_{+1}s[E]$. But if the reaction $E + X \rightleftharpoons EX$ is an equilibrium with dissociation constant K_x, E can be regarded as forming a fraction $K_x/(K_x + x)$ of the concentration of the composite species $\{E + EX\}$, i.e. $[E] = \{[E] + [EX]\}K_x/(K_x + x)$. So the rate of the reaction $E \rightarrow ES$ can be written in terms of the composite species as $k_{+1}K_x s\{[E] + [EX]\}/(K_x + x)$. In other words, if E is regarded as a component of $\{E + EX\}$, the rate constant for the conversion of the composite species $\{E + EX\}$ into ES is $k_{+1}K_x/(K_x + x)$, that is, the original rate constant $k_{+1}s$ multiplied by a weight $K_x/(K_x + x)$ that defines the proportion of the composite species capable of reacting. The conversion of EX into EXS can be treated in exactly the same way, so that the rate constant $k'_{+1}s$ for conversion of EX is multiplied by the weight $x/(K_x + x)$ so that it becomes the rate constant for conversion of $\{E + EX\}$ into EXS. Similarly, because the reaction $ES + X \rightleftharpoons EXS$ is also regarded as an equilibrium, $\{ES + EXS\}$ forms a second composite species, and rate constants for reactions away from it can be defined in the same way. Eventually, the mechanism can be expressed as

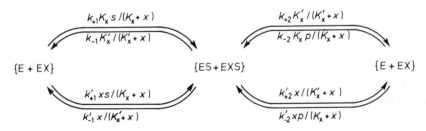

Notice that although the individual rate constants assume a more complicated appearance as a result of this treatment, the derivation of the rate equation is greatly simplified because far less algebra is required. Because there are now only two species in the mechanism, all of the reactions are parallel steps and can be added, by rule 1 of Section 4.2.

One can apply Cha's method to any mechanism in which there are steps at equilibrium. In general, any number of species in equilibrium with one another can be treated as one species, and each rate constant k_i for a component is reduced to $k_i f_i$ for the composite species, where the weight f_i represents the fraction of the component in the mixture. After this treatment it is usually possible to add parallel steps, as in the example above.

This type of simplification is particularly useful in the analysis of pH dependence (Chapter 7) and in the analysis of mechanisms with parallel pathways. In the latter case, it is often convenient to treat the alternative pathways as equilibria and the compulsory pathways as slow steps. Equations derived in this way are commonly in accordance with experiment, but *this does not prove that the underlying assumptions are correct*: it is quite possible for there to be additional terms in the rigorous steady-state equation that are numerically significant and yet virtually impossible to detect, because of near proportionality to other terms in the equation over any reasonable experimental range. Gulbinsky and Cleland (1968) have described an example of this.

4.4 Analysing mechanisms by inspection

Once one is thoroughly conversant with the King–Altman method, it is often possible to reach important conclusions about the rate equation for a mechanism without having to derive it in detail, simply by inspecting the master pattern carefully.

It is an important characteristic of the King–Altman method that every pattern generates a positive term, and that every term appears in the denominator of the rate equation. As there are no negative terms, no terms can cancel by subtraction, and so every term for which a pattern exists must appear in the rate equation. The only exception to this rule is that sometimes the numerator and denominator share a common factor that can be cancelled by division, but this normally happens only if the rate constants are related to one another, as in the following mechanism for an enzyme with two independent and identical active sites:

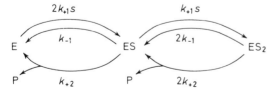

In this case, the rate constants for the second site are the same as those for the first apart from statistical factors, and the rate equation is

$$v = \frac{4k_{+1}k_{+2}e_0 s(k_{-1} + k_{+2}) + 4k_{+1}k_{+2}e_0 s^2}{2(k_{-1} + k_{+2})^2 + 4k_{+1}s(k_{-1} + k_{+2}) + 2k_{+1}^2 s^2}$$

This equation apparently contains terms in s^2, and so one might suppose it to predict deviations from the Michaelis–Menten equation. In fact, however, the numerator and denominator share a common factor, $2(k_{-1} + k_{+2} + k_{+1}s)$, and when this is cancelled the equation simplifies to the Michaelis–Menten equation:

$$v = \frac{2k_{+2}e_0 s}{[(k_{-1} + k_{+2})/k_{+1}] + s}$$

In mechanisms in which there are no relationships between the rate constants other than those required by thermodynamics, it is usually safe to assume that cancellation between numerator and denominator will not be possible, so that any term for which a pattern exists must appear in the rate equation. For example, consider the general modifier mechanism of Botts and Morales (1953) (without assuming any steps to be at equilibrium):

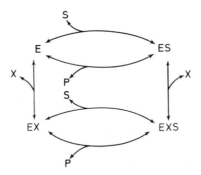

If one wished to confirm that the rate equation for this mechanism contained terms in s^2, one could do so without deriving it by noting that there are two patterns that give rise to terms in s^2:

Apart from the value of inspection for considering individual mechanisms, it can be used to reach important and far-reaching conclusions about mechanisms in general. Consider, for example, the fact that dead-end reactions can be treated as equilibria in the steady state. The reason why most reversible steps in a mechanism cannot be treated as equilibria is that any net flux through a step must unbalance any equilibrium that might otherwise exist. However, in a *dead-end reaction*, i.e. one connected to the rest of the mechanism at one point only, there is no net flux, and thus nothing to prevent equilibrium from being maintained. A dead-end reaction can be introduced into the mechanism discussed in Section 4.3 by supposing that B can bind to EQ to produce a dead-end complex EBQ:

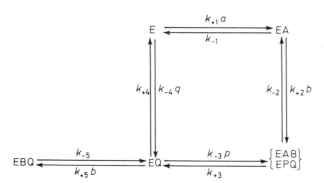

For every species except EBQ, every King–Altman pattern must contain the reaction EBQ → EQ, and so the expression must be the same as before, multiplied by k_{-5}; conversely, every pattern for EBQ must contain the reaction EQ → EBQ and so the expression for EBQ must be the original expression for EQ multiplied by $k_{+5} b$. Thus, $[\text{EBQ}]/[\text{EQ}] = k_{+5} b/k_{-5}$ and the reaction EQ + B ⇌ EBQ is at equilibrium. The rate equation for the mechanism with dead-end inhibition is therefore identical to the rate equation without such inhibition, except that the terms for EQ in the denominator must be multiplied by $(1 + k_{+5} b/k_{-5})$.

Problems

4.1 Derive a rate equation for the following mechanism, treating { EA + E′P} as a single species, and { E′B + EQ} as a single species:

E + A ⇌ EA ⇌ E′P ⇌ E′ + P

E′ + B ⇌ E′B ⇌ EQ ⇌ E + Q

Without carrying out a complete derivation, write down a rate equation for the case where B can bind to E to give a dead-end complex EB.

4.2 Consider an enzyme that catalyses two reactions simultaneously, A + B ⇌ P + Q and A + B′ ⇌ P′ + Q, where A and Q are common to the two reactions but B, B', P and P' are not. Assume that both reactions proceed by the mechanism used as an example in Section 4.1. Draw a master pattern for the system and find all valid King–Altman patterns.

In this system the rate equation for v defined as dp/dt is not the same as that for v defined as $-da/dt$. Why not?

4.3   The mechanism defined in Problem 4.1 above is an example of a general class of mechanisms known as *substituted-enzyme mechanisms*. Show that, for all mechanisms that contain at least two enzyme forms (E and E' in this example) which do not react in unimolecular reactions, every term in the rate equation must contain at least one reactant concentration, i.e. there is no constant term in the equation.

An exercise in the use of Cha's method is given in Problem 7.4 at the end of Chapter 7 (p. 145).

Chapter 5

Inhibitors and activators

5.1 Reversible and irreversible inhibitors

Compounds that decrease the rate of an enzyme-catalysed reaction when present in the reaction mixture are called *inhibitors*. Inhibition can arise in a wide variety of ways, however, and there are many different types of inhibitor. One class that I shall mention only briefly is that of *irreversible inhibitors* or *catalytic poisons*. These are compounds that combine with the enzyme in such a way as to decrease its activity to zero. Many enzymes are poisoned by trace amounts of heavy-metal ions, and for this reason it is common practice to carry out kinetic studies in the presence of complexing agents, such as ethylenediamine tetraacetate. This is particularly important in the purification of enzymes: in crude preparations, the total protein concentration is high and the many protein impurities sequester almost all of the metal ions that may be present, but the purer an enzyme becomes, the less it is protected by other proteins and the more important it is to add alternative sequestering agents. Irreversible inhibitors can also be used in a positive way. For example, poisoning by mercury(II) compounds has often been used to implicate sulphydryl groups in the catalytic activity of enzymes.

Careful measurements of the *rate* of loss of activity in irreversible inhibition sometimes yields information that is simpler, and hence more readily interpretable, than that obtained in conventional enzyme kinetic experiments. The theory commonly applied in this type of study is one given by Kitz and Wilson (1962), who were concerned with the effects of various compounds on acetylcholinesterase. They made essentially the same assumptions as those of Michaelis and Menten (1913) for the catalytic process, i.e. they assumed that the inactivation of an enzyme E by an inhibitor I would proceed through an intermediate EI that was in equilibrium with E and I throughout the process:

$$\text{E} + \text{I} \underset{K_i}{\rightleftharpoons} \text{EI} \xrightarrow{k_{+2}} \text{E}' \text{ (inactive)}$$

$$e_t - x \quad\quad i \quad\quad x$$

where e_t represents the total concentration of active enzyme, i.e. including E and EI but not E'. With these assumptions the concentration of EI at any time is equal to $(e_t - x)i/K_i$ and the rate of inactivation v is given by

(margin note: ? should be $K_i/i \rightarrow$**)**

(margin note near equation: ✗ ok**)**

$$v = k_{+2}x = k_{+2}e_t/(1 + i/K_i)$$

If the inhibitor is in sufficient excess for i to be essentially constant the loss of activity is a pseudo-first-order process and analysis by the methods of Section 1.5 would give an apparent first-order rate constant $k^{app} = k_{+2}/(1 + i/K_i)$. As this is a linear function of i, one can estimate k_{+2} and K_i from measurements of k^{app} at different values of i.

The remainder of this chapter will be concerned with *reversible* inhibitors. These are compounds that form dynamic complexes with the enzyme which have catalytic properties different from those of the uncombined enzyme. The inhibited enzyme may have an increased K_m value (*competitive inhibition*), a decreased V value (*pure noncompetitive inhibition*), V and K_m decreased in constant ratio (*uncompetitive inhibition*), or some combination of these effects (*mixed inhibition*). The simplest kinds of inhibition arise when the form of the enzyme with inhibitor bound to it retains no catalytic activity. These may be called types of *complete inhibition*, but are more often called types of *linear inhibition* because they give rise to linear plots of the apparent values of K_m/V and $1/V$ against the inhibitor concentration. A more complex class of inhibition in which the enzyme retains some activity when the inhibitor is bound to it is called *partial inhibition*, from its mechanistic character, or *hyperbolic inhibition*, from the shapes of the plots that it causes. In principle, both linear and hyperbolic inhibitors can be classified into competitive, uncompetitive and mixed types; but in practice these latter terms are usually assumed to imply linear inhibition unless otherwise stated.

5.2 Competitive inhibition

In the reaction catalysed by succinate dehydrogenase, succinate is oxidized to fumarate:

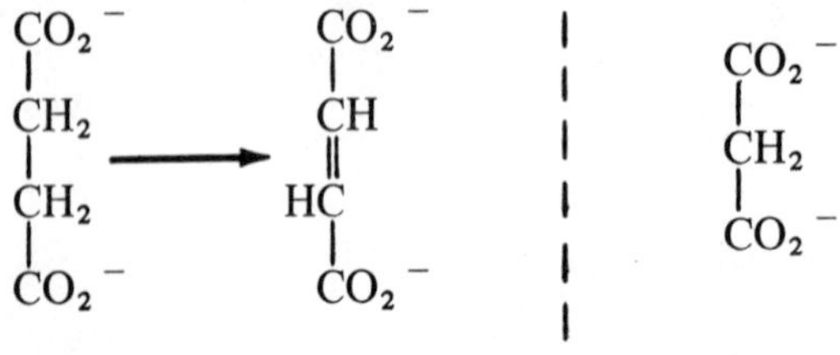

$$\begin{array}{ccc}
\text{CO}_2^- & \text{CO}_2^- & \text{CO}_2^- \\
| & | & | \\
\text{CH}_2 & \text{CH} & \text{CH}_2 \\
| & \| & | \\
\text{CH}_2 & \text{HC} & \text{CO}_2^- \\
| & | & \\
\text{CO}_2^- & \text{CO}_2^- & \\
\text{succinate} & \text{fumarate} & \text{malonate}
\end{array}$$

As this is a reaction of the dimethylene group it plainly cannot occur with malonate, which does not possess a dimethylene group. Malonate is otherwise very similar in structure to succinate, however, and it is not surprising, therefore, that it can bind to the substrate-binding site of succinate dehydrogenase to give an abortive complex that is incapable of reacting. This is an example of the commonest type of inhibition, known as *competitive inhibition*, because the substrate and inhibitor compete for the same site. The mechanism may be represented in general terms as follows:

$$
\begin{array}{c}
\text{EI} \\
\Big\updownarrow K_i \\
\text{E} + \text{S} \rightleftharpoons \text{ES} \rightarrow \text{E} + \text{P}
\end{array}
$$

In this scheme, EI is a dead-end complex, as it can break down only by returning to E + I. Consequently (cf. Section 4.4), its concentration is given by a true equilibrium constant, $K_i = [\text{E}][\text{I}]/[\text{EI}]$, which is termed the *inhibition constant*. In many of the more complex types of inhibition, including most types of product inhibition, the inhibition constant cannot be treated as a true equilibrium constant because the enzyme–inhibitor complex is not a dead-end complex.

The defining equation for linear competitive inhibition, which applies not only to the above mechanism but to some others as well, is

$$
v = \frac{Vs}{K_m(1 + i/K_i) + s} \tag{5.1}
$$

in which i is the free-inhibitor concentration and V and K_m have their usual meanings. The equation is of the form of the Michaelis–Menten equation, i.e. it can be written as

$$
v = \frac{V^{app} s}{K_m^{app} + s}
$$

where V^{app} and K_m^{app} are the *apparent* values of V and K_m and are given by

$$
V^{app} = V
$$

$$
K_m^{app} = K_m(1 + i/K_i)
$$

$$
V^{app}/K_m^{app} = \frac{V/K_m}{1 + i/K_i}
$$

Hence the effect of a competitive inhibitor is to increase the apparent value of K_m by the factor $(1 + i/K_i)$, to decrease that of V/K_m by the same factor, and to leave V unchanged. I mention V/K_m explicitly here because in most situations (though not this one) its behaviour is simpler than that of K_m.

5.3 Mixed inhibition

Most elementary accounts of inhibition discuss two types of inhibition only, competitive inhibition and *non-competitive inhibition*. Competitive inhibition is of genuine importance, but non-competitive inhibition is a phenomenon that rarely occurs in practice and it need not be considered in detail here. It arose originally because the earliest students of inhibition, Michaelis and his collaborators, assumed that certain inhibitors acted by decreasing the apparent value of V, but had no effect on K_m. This effect would be an obvious alternative to competitive inhibition, and was termed 'non-competitive inhibition'. It is difficult to imagine a reasonable explanation of such effects, however: one would have to assume that the inhibitor interfered with the catalytic properties of the enzyme, but that it had no effect on the binding of substrate. This might be possible for very small inhibitors, such as protons or metal ions, but seems most unlikely otherwise. In fact, non-competitive inhibition or activation by protons is common and there are several instances of non-competitive inhibition by heavy-metal ions. Non-competitive inhibition by other species is very rare, however, and most of the commonly quoted examples, such as the inhibition of invertase by α-glucose (Nelson and Anderson, 1926) and the inhibition of arginase by various compounds (Hunter and Downs, 1945), prove, on re-examination of the original data, to be examples of *mixed inhibition*. In general, it is best to regard non-competitive inhibition as a special, and not very interesting, case of mixed inhibition.

Linear mixed inhibition occurs if both V^{app} and V^{app}/K_m^{app} vary with the inhibitor concentration according to the equations

$$V^{app} = \frac{V}{1 + i/K_i'} \tag{5.2}$$

$$K_m^{app} = \frac{K_m (1 + i/K_i)}{1 + i/K_i'} \tag{5.3}$$

and

$$V^{app}/K_m^{app} = \frac{V/K_m}{1 + i/K_i} \tag{5.4}$$

The simplest formal mechanism for mixed inhibition is

$$
\begin{array}{ccc}
\text{EI} & & \text{EIS} \\
\Big\updownarrow K_i & & \Big\updownarrow K'_i \\
\text{E} + \text{S} \underset{k_{-1}}{\overset{k_{+1}}{\rightleftharpoons}} \text{ES} & \xrightarrow{k_{+2}} & \text{E} + \text{P}
\end{array}
$$

The inhibitor can bind both to the free enzyme to give a complex EI with dissociation constant K_i, and to the ES complex to give an unreactive EIS complex with dissociation constant K'_i. As shown in this scheme, both inhibitor-binding reactions are dead-end reactions and are therefore equilibria (*see* Section 4.4). As both EI and EIS exist, however, it is difficult to see why S should not bind directly to EI to give EIS. If this reaction is included in the mechanism the rate equation becomes much more complicated, because terms in s^2 and i^2 appear in it. These terms cancel only if all of the binding reactions are equilibria, i.e. if the substrate- and product-release steps are all fast compared with the reaction that converts ES into products. In practice, however, the predicted deviations from simple kinetics are difficult to detect experimentally, and one cannot use adherence to simple kinetics as evidence that K_m, K_i and K'_i are true dissociation constants.

Although it is formally convenient to define mixed inhibition in terms of the scheme shown above, it actually occurs mainly as an important case of product inhibition. If a product is released in a step that generates an enzyme species other than that to which the substrate binds, product inhibition is expected to be in accordance with equations 5.2–5.4. This conclusion does not depend on any equilibrium assumptions, i.e. it is a necessary consequence of the steady-state treatment, as can readily be shown by the methods of Chapter 4. The simplest of many mechanisms of this type is one in which the product is released in the second of three steps:

$$
\text{E} + \text{S} \underset{k_{-1}}{\overset{k_{+1}}{\rightleftharpoons}} \text{ES} \underset{k_{-2}}{\overset{k_{+2}}{\rightleftharpoons}} \text{E}' + \text{P}
$$

$$
k_{+3}
$$

More complex examples abound in reactions that involve more than one substrate or product, as will be seen in Chapter 6. In these cases, identification of K_i and K'_i with dissociation constants is not very useful. Even in this simple example, $K_i = (k_{-1} + k_{+2})k_{+3}/k_{-1}k_{-2}$ and $K'_i = (k_{+2} + k_{+3})/k_{-2}$, neither of which is an equilibrium constant except in special cases, such as $k_{+3} \ll k_{+2}$.

Because of the rareness of non-competitive inhibition, some enzymologists have generalized the term to embrace mixed inhibition. There seems to be no advantage in doing this, and it is a most unfortunate development, as it has added ambiguity to an already confused nomenclature. To avoid this ambiguity one must refer to non-competitive inhibition in the traditional sense as *pure* non-competitive inhibition, on the rare occasions when one wants to refer to it at all.

5.4 Uncompetitive inhibition

The last of the simple types of inhibition to be considered is known, rather unhelpfully, as *uncompetitive inhibition*. It is characterized by equal effects on V and K_m, but no effect on V/K_m :

$$V^{\mathrm{app}} = \frac{V}{1 + i/K_i'}$$

$$K_m^{\mathrm{app}} = \frac{K_m}{1 + i/K_i'}$$

$$V^{\mathrm{app}}/K_m^{\mathrm{app}} = V/K_m$$

Comparison of these equations with equations 5.2–5.4 shows that uncompetitive inhibition is a limiting case of mixed inhibition in which K_i approaches infinity (i.e. i/K_i is negligible at all values of i and hence disappears from equations 5.3 and 5.4). It is thus the converse of competitive inhibition, which is the other limiting case of mixed inhibition in which K_i' approaches infinity.

Uncompetitive inhibition is also, at least in principle, the mechanistic converse of competitive inhibition, because it is predicted for mechanisms in which inhibitor binds only to the ES complex and not to the free enzyme. Such mechanisms are not particularly plausible, however, and uncompetitive inhibition occurs almost exclusively as a type of product inhibition that is common in reactions with several substrates and products.

5.5 Plotting inhibition results

The properties of the various types of linear inhibition discussed in the preceding three sections are summarized in *Table 5.1*. These properties are easy to memorize as long as the following points are noted:

(1)   The two limiting cases are competitive and uncompetitive inhibition; pure non-competitive inhibition is simply a special case of mixed inhibition in which K_i and K_i' are equal.

(2)   The effects of inhibitors on V^{app} and $V^{\mathrm{app}}/K_m^{\mathrm{app}}$ are simple and regular; if they are decreased at all by the inhibitor they

TABLE 5.1   Characteristics of linear inhibitors

Type of inhibition	V^{app}	V^{app}/K_m^{app}	K_m^{app}
Competitive	V	$\dfrac{V/K_m}{1 + i/K_i}$	$K_m(1 + i/K_i)$
Mixed	$\dfrac{V}{1 + i/K_i'}$	$\dfrac{V/K_m}{1 + i/K_i}$	$\dfrac{K_m(1 + i/K_i)}{1 + i/K_i'}$
Pure non-competitive†	$\dfrac{V}{1 + i/K_i'}$	$\dfrac{V/K_m}{1 + i/K_i}$	K_m
Uncompetitive	$\dfrac{V}{1 + i/K_i'}$	V/K_m	$\dfrac{K_m}{1 + i/K_i'}$

†As pure non-competitive inhibition is simply a special case of mixed inhibition with $K_i = K_i'$, it is not strictly necessary to use *both* symbols in this line of the table, but this is done to preserve the regularity of the columns for V^{app} and V^{app}/K_m^{app}.

are decreased by factors of $(1 + i/K_i')$ and $(1 + i/K_i)$ respectively.

(3) The effects of inhibitors on K_m^{app} are complex and confusing; they are most easily remembered by regarding K_m^{app} as the ratio of V^{app} and V^{app}/K_m^{app}, rather than as a parameter in its own right.

Any of the plots described in Section 2.5 can be used to diagnose the type of inhibition, as they all provide estimates of the apparent values of the kinetic parameters. For example, if plots of s/v against s are made at several values of i, the intercept on the ordinate (K_m^{app}/V^{app}) varies with i if there is a competitive component in the inhibition, and the slope ($1/V^{app}$) varies with i if there is an uncompetitive component. Alternatively, if direct linear plots of V^{app} against K_m^{app} are made at each value of i, the common intersection point shifts in a direction that is characteristic of the type of inhibition: for competitive inhibition, the shift is to the right; for uncompetitive inhibition, it is directly towards the origin; and for mixed inhibition, it is intermediate between these extremes. Examples of the various possibilities are shown schematically in *Figure 5.1*, and an experimental example of competitive inhibition is shown in *Figure 5.2*.

Other plots are needed for determining the actual values of K_i and K_i'. The simplest approach is to estimate the apparent kinetic constants at several values of i, by the methods of Section 2.5, and to plot K_m^{app}/V^{app} and $1/V^{app}$ against i. In each case, a straight line is obtained, and the intercept on the i axis gives $-K_i$ if K_m^{app}/V^{app} is plotted, or $-K_i'$ if $1/V^{app}$ is plotted. Now, it may seem more natural

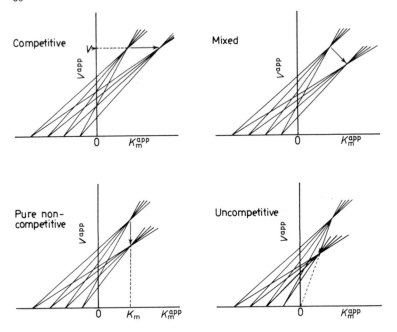

Figure 5.1 Effect of various types of inhibition on the location of the common intersection point (K_m^{app}, V^{app}) of the direct linear plot (cf. *Figure 2.6*)

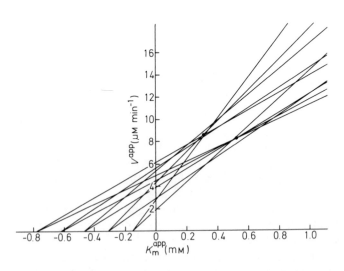

Figure 5.2 Direct linear plots showing competitive inhibition. The figure is taken from Eisenthal and Cornish-Bowden (1974) and shows data of Cornish-Bowden (1967) for the inhibition of the pepsin-catalysed hydrolysis of N-acetyl-3,5-dinitro-L-tyrosyl-L-phenyl-alanine by acetyl-L-phenylalanyl-D-phenylalanine. At each substrate concentration, the higher rate was observed in the absence of inhibitor, the lower in the presence of 0.525 mM inhibitor

to determine K_i by plotting K_m^{app} rather than K_m^{app}/V^{app} against i, but this is not advisable for the following two reasons. It is valid only if the inhibition is competitive, and it gives a curve rather than a straight line if the inhibition is mixed; it is also much less accurate, even if the inhibition is competitive, because K_m^{app} can never be estimated as precisely as can K_m^{app}/V^{app}.

Another method of estimating K_i, introduced by Dixon (1953b), is also in common use. By taking reciprocals of both sides of the full equation for mixed inhibition,

$$v = \frac{Vs}{K_m(1 + i/K_i) + s(1 + i/K_i')} \tag{5.5}$$

we can obtain the following two equations, in which the subscripts 1 and 2 indicate measurements at two different values of s:

$$\frac{1}{v_1} = \frac{K_m + s_1}{Vs_1} + \frac{(K_m/K_i + s_1/K_i')i}{Vs_1}$$

$$\frac{1}{v_2} = \frac{K_m + s_2}{Vs_2} + \frac{(K_m/K_i + s_2/K_i')i}{Vs_2}$$

Both of these equations indicate that a plot of $1/v$ against i at a constant value of s is a straight line. If two such lines are drawn, from measurements at two different s values, the point of intersection can be found by setting $1/v_1$ equal to $1/v_2$. This gives

$$\frac{K_m + s_1}{Vs_1} + \frac{(K_m/K_i + s_1/K_i')i}{Vs_1} = \frac{K_m + s_2}{Vs_2} + \frac{(K_m/K_i + s_2/K_i')i}{Vs_2}$$

Hence

$$\frac{K_m}{V}\left(\frac{1}{s_1} - \frac{1}{s_2}\right)\left(1 + \frac{i}{K_i}\right) = 0$$

and so $i = -K_i$ at the point of intersection. In principle, if several lines are drawn at different s values, they should all intersect at a common point; in practice, experimental error will usually ensure some variation. Notice that in the derivation the terms that contained K_i' cancelled out. Consequently the plot provides the value of K_i regardless of the value of K_i', that is, regardless of whether the inhibition is competitive, mixed or pure non-competitive. By the same token the plot provides no information about the value of K_i', and in the case of uncompetitive inhibition, in which K_i is infinite, it generates parallel lines.

Although the Dixon plot does not provide the value of K_i', the uncompetitive inhibition constant, an exactly similar derivation shows

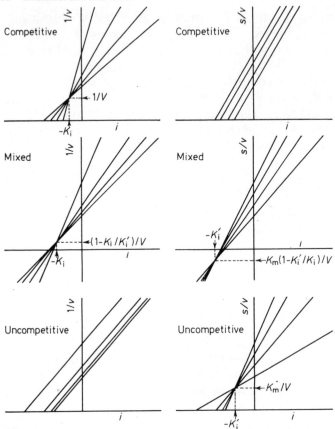

Figure 5.3 Determination of (a) K_i, from plots of $1/v$ against i at various s values (Dixon, 1953b), and (b) K_i' from plots of s/v against i at various s values (Cornish-Bowden, 1974). In the case of mixed inhibition, the point of intersection can be above the axis in the first plot and below it in the second, or vice versa, or, if $K_i = K_i'$, on the axis in both plots

that this value can be found by plotting s/v against i at several s values (Cornish-Bowden, 1974). In this case, a different set of straight lines is obtained that intersect at a point where $i = -K_i'$. Both types of plot are shown schematically for the various types of inhibition in *Figure 5.3*, and an experimental example showing competitive inhibition is shown in *Figure 5.4*.

5.6 Inhibition by a competing substrate

It is not particularly common practice to carry out kinetic experiments in which two or more substrates compete for the same enzyme[†]

[†]This situation should be carefully distinguished from the case of enzymes that *require* two or more substrates for the reaction to be complete, such as hexokinase, which was discussed in Section 3.3. In that example, glucose and ATP are not competing substrates because both must be present for a complete reac-

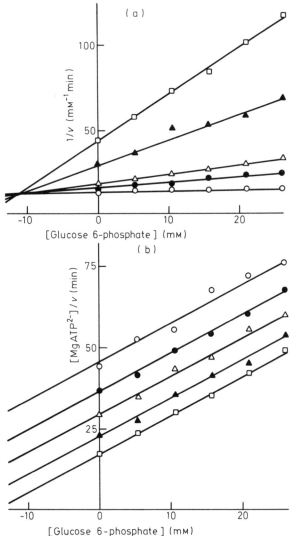

Figure 5.4 Inhibition of glucokinase by glucose 6-phosphate at various MgATP²⁻ concentrations (Storer and Cornish-Bowden, 1977), plotted as in *Figure 5.3*. The different symbols represent different concentrations of MgATP²⁻. The set of intersecting lines in (a) combined with the set of parallel lines in (b) indicate competitive inhibition with $K_i = 11.5$ mM

tion to be possible. Hexokinase will, however, accept fructose and other hexoses as alternatives to glucose as substrates. If glucose and fructose are simultaneously available to the enzyme, as certainly occurs in the living cell, they are then competing substrates.

Inhibition by a competing substrate should also be distinguished from *substrate inhibition*: this term is reserved for the inhibition that occurs as a result of binding of an additional molecule of the substrate of the reaction being catalysed, to give an inactive complex. It is discussed in Section 5.11.

because this tends to make the analysis complicated without providing much more information than would be obtained by studying the substrates separately. Nonetheless, most enzymes are not perfectly specific for a single substrate and it often happens in the cell that several possible substrates are simultaneously available to one enzyme. Thus for a realistic appraisal of physiological conditions one ought to examine the kinetics of competing substrates. The simplest case occurs when there are two competing substrates, each of which gives rise to Michaelis–Menten kinetics when studied in absence of the other:

$$E + A \underset{k_{-11}}{\overset{k_{+11}}{\rightleftharpoons}} EA \overset{k_{+12}}{\longrightarrow} E + P$$

$$E + B \underset{k_{-21}}{\overset{k_{+21}}{\rightleftharpoons}} EB \overset{k_{+22}}{\longrightarrow} E + Q$$

This is a somewhat simplified version of the mechanism for competitive substrates used as an illustration in Section 4.2. It generates the following pair of rate equations:

$$v_1 = \frac{dp}{dt} = \frac{V^A a}{K_m^A (1 + b/K_m^B) + a} \tag{5.6}$$

$$v_2 = \frac{dq}{dt} = \frac{V^B b}{K_m^B (1 + a/K_m^A) + b} \tag{5.7}$$

in which $V^A = k_{+12} e_0$ and $V^B = k_{+22} e_0$ are the maximum velocities and $K_m^A = (k_{-11} + k_{+12})/k_{+11}$ and $K_m^B = (k_{-21} + k_{+22})/k_{+21}$ are the Michaelis constants of the two reactions in isolation. There are two points to note about these equations. First, they are precisely of the form of equation 5.1 for competitive inhibition, and so measurement of the 'inhibition constant' for a competing substrate by treating it as if it were an inhibitor must actually yield the Michaelis constant. This may be seen in the example shown in *Table 5.2*, in which K_m has been measured for several poor substrates of fumarase both directly and in a competitive reaction. In each case the values of K_m measured in the two different ways agree to within experimental error.

The second point to note about equations 5.6 and 5.7, which is also illustrated by the data in *Table 5.2*, is that they permit a rigorous definition of what is meant by enzyme *specificity*. Consider the parameters for fluorofumarate and for fumarate. The value of k_{cat} for fluorofumarate is about three times greater than that for fumarate; thus at high concentrations fluorofumarate appears to be a better

TABLE 5.2 Kinetic parameters for substrates of fumarase

The data refer to measurements at 25 °C in buffer of pH 7.3. Values of k_{cat} (i.e. V/e_0) and K_m were measured in conventional kinetic experiments. Values in the column labelled 'K_i' are K_m values for the poor substrates measured by treating them as competitive inhibitors of the reaction with fumarate as substrate. The table is adapted from Teipel, Hass and Hill (1968)

Substrate	k_{cat} (s^{-1})	K_m (mM)	'K_i' (mM)	k_{cat}/K_m (s^{-1} mM^{-1})
Fluorofumarate	2 700	0.027	–	100 000
Fumarate	800	0.005	–	160 000
Chlorofumarate	20	0.11	0.10	180
Bromofumarate	2.8	0.11	0.15	25
Iodofumarate	0.043	0.12	0.10	0.36
Mesaconate	0.023	0.51	0.49	0.047
L-Tartrate	0.93	1.3	1.0	0.72

substrate than fumarate, if the two reactions are considered in isolation. The reverse is true at low concentrations, however, because k_{cat}/K_m is about 60% greater than for fluorofumarate. Which of these results is the more fundamental, that is, which is truly the more specific substrate? The question may seem to be one of definitions, but a clear and satisfying answer emerges when one realizes that it is artificial to consider each substrate in isolation from the other. In most physiological discussions of specificity one ought to consider the proportion of reaction utilizing each substrate when they are mixed together. This can be determined by dividing equation 5.6 by equation 5.7:

$$\frac{v_1}{v_2} = \frac{dp}{dq} = \frac{k_{+11}k_{+12}a}{k_{+21}k_{+22}b} = \frac{(V^A/K_m^A)a}{(V^B/K_m^B)b} = \frac{(k_{cat}^A/K_m^A)a}{(k_{cat}^B/K_m^B)b}$$

Thus it is the ratio of k_{cat}/K_m values for the two substrates that determines the ratio of rates of the competing reactions when the substrates are mixed together. It follows that in an equimolar mixture of fumarate and fluorofumarate, *at any concentration*, the rate of the fumarase-catalysed hydration of fumarate is 60% faster than the hydration of fluorofumarate. Clearly therefore fumarate is the more specific substrate, and k_{cat}/K_m is the fundamental quantity to be considered in discussions of specificity, regardless of concentration.

5.7 Activation

Any discussion of the *activation* of enzyme-catalysed reactions is likely to be complicated by the fact that the term has been used in enzymology with several disparate meanings. In this book I use it for the converse of reversible inhibition: an activator is a species that combines with an enzyme so as to increase its activity, without itself undergoing a net change in the reaction. Other processes that are sometimes called activation are the following:

(1)   Several enzymes, mainly extracellular catabolic enzymes such as pepsin, are secreted as inactive precursors or *zymogens*, pepsinogen in the case of pepsin, which are subsequently converted into the active enzyme by partial proteolysis. This process is sometimes called 'zymogen activation'.

(2)   Several enzymes important in metabolic regulation, such as glycogen phosphorylase, exist in the cell in active and inactive states, the two differing by the presence or absence of a phosphate group. The conversion between the two states requires two separate reactions, transfer of a phosphate group from ATP in one direction and removal of the phosphate group by hydrolysis in the other, and neither process corresponds to activation or inhibition in the dynamic sense used in this book.

(3)   Many reactions are said to be 'activated' by metal ions when the truth is that a metal ion forms part of the substrate. For example, nearly all ATP-dependent kinases are 'activated' by Mg^{2+}, not because of the effect of Mg^{2+} on the enzyme itself, but because the true substrate is the ionic species $MgATP^{2-}$, not ATP^{4-}, the predominant metal-free form at physiological pH. For example, although rat-liver glucokinase uses $MgATP^{2-}$ as substrate, free Mg^{2+} is actually an inhibitor, not an activator, as indeed ATP^{4-} is another inhibitor, not a substrate (Storer and Cornish-Bowden, 1977). Although this sort of confusion is understandable if one regards ATP as the reactant in reactions that actually involve $MgATP^{2-}$, it is best to avoid it by expressing results in terms of the concentrations of the actual species involved and restricting the term 'activation' to effects on the enzyme. Because of the great importance of $MgATP^{2-}$ in metabolic reactions, I have discussed methods of controlling its concentration in Section 3.10.

The simplest kind of true activation is *compulsory activation*, in which the free enzyme without activator bound to it has no activity and does not bind substrate. This may be represented by the scheme

$$S + EX \underset{k'_{-1}}{\overset{k'_{+1}}{\rightleftharpoons}} EXS \xrightarrow{k'_{+2}} EX + P$$

$$\updownarrow K_X$$

$$E + X$$

in which the activator is represented as X. This scheme is similar to that for competitive inhibition, and it generates a rate equation of similar form:

$$v = \frac{V's}{K'_m (1 + K_x/x) + s}$$

in which $V' = k'_{+2} e_0$ and $K'_m = (k'_{-1} + k'_{+2})/k'_{+1}$ are the maximum velocity and Michaelis constant respectively for the activated enzyme EX, and x is the concentration of X.

This equation differs from that for competitive inhibition (equation 5.1) by having i/K_i replaced with K_x/x. Thus the rate is zero in the absence of activator, as one would expect from the mechanism. Despite the formal similarity between competitive inhibition and compulsory activation, however, there is an important difference in plausibility. Because a competitive inhibitor is conceived to bind at the same site on the enzyme as the substrate, it is easy to imagine that they cannot bind simultaneously; hence one can readily understand why linear competitive inhibition is a common phenomenon. But it is less easy to visualize an enzyme that cannot bind substrate at all in the absence of activator. Moreover, one of the commonest activators is the proton, which has no bulk and so no steric effect. Consequently simple compulsory activation is much less frequently encountered than its counterpart in inhibition; it is useful mainly as a simple introduction to the more complex kinds of activation, which unfortunately require correspondingly complex rate equations. The simplest type of activation that is at all plausible is the counterpart of mixed inhibition:

$$\text{S} + \text{EX} \underset{k'_{-1}}{\overset{k'_{+1}}{\rightleftharpoons}} \text{EXS} \overset{k'_{+2}}{\longrightarrow} \text{EX} + \text{P}$$
$$\Big\updownarrow K_x \qquad \Big\updownarrow K'_x$$
$$\text{E} + \text{X} \qquad \text{ES} + \text{X}$$

In this case the activator is not required for substrate binding, but only for catalysis. The rate equation is analogous to that for mixed inhibition:

$$v = \frac{V's}{K'_m (1 + K_x/x) + s(1 + K'_x/x)} \qquad (5.8)$$

and provides analogous expressions for the apparent parameters (cf. equations 5.2–5.4):

$$V^{\text{app}} = \frac{V'}{1 + K'_x/x}$$

$$K_m^{app} = \frac{K_m'(1 + K_x/x)}{1 + K_x'/x}$$

$$V^{app}/K_m^{app} = \frac{V'/K_m'}{1 + K_x/x}$$

Essentially the same plots and methods can be used to investigate the type of activation as are used in linear inhibition (Section 5.5), replacing i throughout by $1/x$ and K_i (or K_i') by $1/K_x$ (or $1/K_x'$). For example, K_x may be determined by a plot analogous to a Dixon plot in which $1/v$ is plotted against $1/x$ at two or more values of s; the abscissa co-ordinate of the point of intersection of the resulting straight lines then gives $-1/K_x$.

5.8 Hyperbolic inhibition and activation

In practice, activators often behave in a more complex way than that suggested by equation 5.8, because there may be some activity in the absence of activator. If this is so, but the rate constants for the two forms of the enzyme are different, we have the scheme

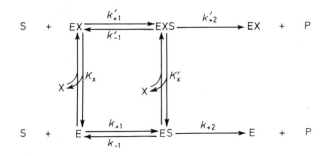

This is the *general modifier mechanism* of Botts and Morales (1953), in which the term *modifier* is used as a general term that embraces both activators and inhibitors. This is because the mechanism is not confined to activation but can also account for more complex types of inhibition than those discussed in the earlier part of this chapter. With this mechanism, plots of $1/V^{app}$ or K_m^{app}/V^{app} against inhibitor or reciprocal activator concentration are not straight lines but rectangular hyperbolas, and so it is a mechanism for *hyperbolic* inhibition and activation. If $k_{+2} > k_{+2}'$ and $k_{+1}k_{+2}/(k_{-1} + k_{+2}) > k_{+1}'k_{+2}'(k_{-1}' + k_{+2}')$, then X is a hyperbolic inhibitor at all substrate concentrations; if the reverse inequalities are obeyed X is a hyperbolic activator at all substrate concentrations; if only one inequality applies, X is an inhibitor in one range of substrate concentrations but an activator elsewhere.

These hyperbolic effects are too complex to justify detailed discussion in an elementary account, but they should not be forgotten completely. The Botts–Morales scheme describes a very plausible mechanism, and it is likely that few examples have been reported, not because the actual number of examples is small, but because the number of examples of inhibition and activation that have been adequately characterized is small. It is not difficult to detect hyperbolic effects qualitatively, though the symptoms are often ignored as unwanted complexities. One must first of all use a wide enough range of inhibitor and activator concentrations to know whether the rate tends to zero at very high inhibitor or very low activator concentration. One should particularly note whether the expected 'linear' plots are actually straight lines or not; any systematic curvature should be checked and if confirmed it is likely to indicate hyperbolic effects.

5.9 Design of inhibition experiments[†]

There are two primary aims in inhibition experiments, to identify the type of inhibition and to estimate the values of the inhibition constants. With a carefully designed experiment it is usually possible to satisfy both of these aims simultaneously. I shall initially assume that the inhibition is linear and that the relationships shown in *Table 5.1* apply, because this is likely to be adequate as a first approximation and it is useful to characterize the simple behaviour before attempting to understand any complexities that may occur.

It is evident from *Table 5.1* that any competitive component in the inhibition will be most pronounced at low concentrations of substrate, because competitive inhibitors decrease V^{app}/K_m^{app}, which characterizes the kinetics at low concentrations of substrate. Conversely, any uncompetitive component will be most noticeable at high substrate concentrations. It is obvious, therefore, that the inhibition can be fully characterized only if it is investigated at both high and low substrate concentrations. Consequently conditions that are ideal for assaying an enzyme may well be unsatisfactory for investigating its response to inhibitors. For example, a simple calculation shows that a competitive inhibitor at a concentration equal to its K_i value decreases the measured rate by less than 10% if $s = 10K_m$;

---

† It may seem illogical to discuss design of inhibition experiments *after* discussing analysis (Section 5.5), as design should obviously precede analysis in a well-planned experiment. However, it is my experience that the most effective way of learning the importance of design is by suffering the difficulties that arise when one tries to analyse the results of an experiment that has been executed without any thought being given to the information it is expected to provide. Moreover, I believe one has to have some knowledge of analytical methods before one can appreciate the principles of design.

although an effect of this size ought to be easily detected in any
careful experiment, it might nonetheless be dismissed as of little consequence if it was not realized that the effect at low substrate concentrations would be much larger.

Just as in a simple experiment without inhibitors it is prudent to
include s values from about $0.5K_m$ to as high as conveniently possible (Section 3.7), so in an inhibition experiment the i values should
extend from about $0.5K_i$ or $0.5K_i'$ (whichever is the smaller) to as
high as possible without making the rate too small to measure accurately. At each i value the s values should be chosen as in Section 3.7,
although relative to K_m^{app} rather than to K_m. This is because it is the
apparent values of the kinetic parameters that characterize the inhibition, not the actual values. A simple example is given in *Table 5.3*.
Note that there is no requirement to use exactly the same set of s
values at each i value, and if one does one will include some measurements that provide little information. Nonetheless, for purposes of
plotting the results it is usually convenient to use many of the same s
values in each experiment. If this is done, one has a reasonable number of data points on every line whether one plots against s as
abscissa (for example in determining apparent Michaelis–Menten
parameters at each i value) or against i as abscissa (for example in
plots of $1/v$ or s/v against i). Accordingly, the s values suggested in
Table 5.3 include a few that would probably not be included if each
line of the table were considered in isolation from the others.

TABLE 5.3   Design of inhibition experiments

The table illustrates the choice of s values required to determine the inhibition constants K_i and K_i' in an experiment where the approximate values are
known to be as follows: $K_m = 1$, $K_i = 2$, $K_i' = 10$, in arbitrary units in each
case. The suggested s values are in the same units as K_m, and are designed to
extend from about $0.5K_m^{app}$ to about $10K_m^{app}$ at each i value.

i	$1 + i/K_i$	$1 + i/K_i'$	K_m^{app}						
							s values		
0	1.0	1.0	1.0	0.5	1	2	5	10	
1	1.5	1.1	1.4	0.5	1	2	5	10	
2	2.0	1.2	1.7		1	2	5	10	20
5	3.5	1.5	2.3		1	2	5	10	20
10	6.0	2.0	3.0		1	2	5	10	20
20	11.0	3.0	3.7			2	5	10	20	50

If these recommendations are followed (as a guide, of course, not
rigidly, because the particular numbers used in *Table 5.3* are unlikely
to apply precisely to any specific example), any hyperbolic character
in the inhibition ought to be obvious without the need for further
experiments. Consequently there is no need to add appreciably to
the remarks at the end of the previous section. The main essential is

to include i values that are high enough and numerous enough to indicate whether the rate approaches zero as the inhibitor approaches saturation or not.

5.10 Non-productive binding

Much of the information that exists about the general properties of enzymes has been obtained from the study of a small group of enzymes, namely the extracellular hydrolytic enzymes, including pepsin, lysozyme, ribonuclease and, most notably, chymotrypsin. These enzymes share various properties that make them eminently suitable for detailed study: they are abundant, easily crystallized, stable, monomeric and can be treated as single-substrate enzymes, as the second substrate is water in each case. But it should not be thought that these are typical properties of enzymes, and all of these enzymes share a further unusual characteristic, one that is a definite disadvantage: their natural substrates are all ill-defined polymers, and as a result they are nearly always studied with unnatural substrates that are much less bulky than the natural ones. However, an enzyme that is capable of binding a polymer is likely to be able to bind a small molecule in many ways. Thus, instead of a single enzyme–substrate complex that breaks down to products, there may be in addition numerous *non-productive complexes* that do not break down. This is illustrated in the following scheme, in which SE represents all non-productive complexes:

$$
\begin{array}{c}
\text{SE} \\[2pt]
\Big\updownarrow K_i \\[2pt]
\text{E} \; + \; \text{S} \;\underset{k_{-1}}{\overset{k_{+1}}{\rightleftharpoons}}\; \text{ES} \;\xrightarrow{k_{+2}}\; \text{E} \; + \; \text{P}
\end{array}
$$

This scheme is the same as that for linear competitive inhibition (Section 5.2) with the inhibitor replaced with substrate, and the rate equation (cf. equation 5.1) is

$$
v = \frac{k_{+2}\,e_0\,s}{\left(\dfrac{k_{-1} + k_{+2}}{k_{+1}}\right)\left(1 + \dfrac{s}{K_i}\right) + s}
$$

If the *expected* values of V and K_m are defined as the values they would have if no non-productive complexes were formed, i.e.

$V^{exp} = k_{+2} e_0$, $K_m^{exp} = (k_{-1} + k_{+2})/k_{+1}$ (cf. 'pH-corrected' constants, Section 7.4), then this equation can be rearranged to give

$$v = \frac{Vs}{K_m + s}$$

where

$$V = \frac{V^{exp}}{1 + K_m^{exp}/K_i}$$

$$K_m = \frac{K_m^{exp}}{1 + K_m^{exp}/K_i}$$

$$V/K_m = V^{exp}/K_m^{exp}$$

Thus the Michaelis–Menten equation is obeyed exactly for this mechanism and so the observed kinetics do not indicate whether non-productive binding is significant or not. Unfortunately, it is often the expected values that are of interest in an experiment, because they refer to the main productive catalytic pathway. Hence the measured values of V and K_m may be less, by an unknown and unmeasurable amount, than the quantities of interest. Only V/K_m gives a correct measure of the catalytic properties of the enzyme.

For highly specific enzymes, plausibility arguments can be used to justify excluding non-productive binding from consideration, but for unspecific enzymes, such as chymotrypsin, comparison of the results for different substrates may sometimes provide evidence of the phenomenon. For example, Ingles and Knowles (1967) measured the rates of hydrolysis of a series of acylchymotrypsins, by measuring k_{cat} values for the chymotrypsin-catalysed hydrolysis of the corresponding p-nitrophenyl esters, in which the 'deacylation' step, i.e. hydrolysis of the acylchymotrypsin intermediate, was known to be rate-limiting. The results (Table 5.4) were somewhat complicated by the fact that the various acyl groups were not equally reactive towards nucleophiles. Ingles and Knowles therefore measured the corresponding rate constants for hydrolysis catalysed by hydroxide ion. On dividing the rate constants for chymotrypsin catalysis by those for base catalysis a most interesting pattern emerged: the order of reactivity of the specific L substrates was exactly reversed with the poor D substrates, i.e. Ac-L-Trp > Ac-L-Phe ≫ Ac-L-Leu ≫ Ac-Gly ≫ Ac-D-Leu ≫ Ac-D-Phe > Ac-D-Trp. The simplest explanation is in terms of non-productive binding: for acyl groups with the correct L configuration, the large hydrophobic sidechains permit tight and rigid binding in the correct mode, largely ruling out non-productive complexes; but for acyl groups with the D configuration,

TABLE 5.4 Non-productive complexes in chymotrypsin catalysis

The table shows data of Ingles and Knowles (1967) for the chymotrypsin-catalysed hydrolysis of the p-nitrophenyl esters of various acetyl amino acids. For these substrates the hydrolysis of the corresponding acetylaminoacyl-chymotrypsins is rate-limiting, and so the measured k_{cat} values are actually first-order rate constants for this hydrolysis reaction. The values are compared with the second-order rate constants k_{OH^-} for base-catalysed hydrolysis of the p-nitrophenyl esters of the corresponding benzyloxycarbonyl amino acids.

Acyl group	k_{cat} (s^{-1})	k_{OH^-} $(M^{-1} s^{-1})$	k_{cat}/k_{OH^-} (M)
Acetyl-L-tryptophanyl-	52	0.16	330
Acetyl-L-phenylalanyl-	95	0.54	150
Acetyl-L-leucyl-	5.0	0.35	14
Acetylglycyl-	0.30	0.51	0.58
Acetyl-D-leucyl-	0.034	0.35	0.097
Acetyl-D-phenylalanyl-	0.015	0.54	0.027
Acetyl-D-tryptophanyl-	0.002 8	0.16	0.018

the same sidechains favour tight and rigid binding in non-productive modes.

Non-productive binding is not usually considered in the context of inhibition; indeed, it is usually not considered at all, but it is plainly a special type of competitive inhibition and it is important to be aware of it when interpreting results for several substrates of an unspecific enzyme. The term *substrate inhibition* is usually reserved for the uncompetitive analogue of non-productive binding, which is considered in the next section.

5.11 Substrate inhibition

For some enzymes it is possible for a second substrate molecule to bind to the enzyme–substrate complex, ES, to produce an inactive complex SES:

$$
E \; + \; S \underset{k_{-1}}{\overset{k_{+1}}{\rightleftharpoons}} ES \; \overset{\substack{SES \\ \big\updownarrow K_{si} \; k_{+2}}}{} \longrightarrow E \; + \; P
$$

This scheme is analogous to that for uncompetitive inhibition (Section 5.4), and gives the following equation for the initial rate:

$$
v = \frac{k_{+2} e_0 s}{\dfrac{k_{-1} + k_{+2}}{k_{+1}} + s(1 + s/K_{si})} = \frac{Vs}{K_m + s + s^2/K_{si}} \tag{5.9}
$$

where V and K_m are defined in the usual way as $k_{+2}e_0$ and $(k_{-1} + k_{+2})/k_{+1}$, respectively. This equation is not of the form of the Michaelis–Menten equation, by virtue of the term in s^2. This term becomes significant only at high substrate concentrations. Hence the rate approaches the Michaelis–Menten value when s is small, but approaches zero instead of V when s is large. The curve of v against s predicted by equation 5.9 is illustrated in *Figure 5.5*, together with

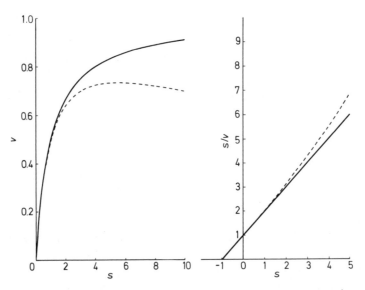

Figure 5.5   Effect of substrate inhibition on plots of v against s and of s/v against s. In both plots, the continuous lines are calculated with $K_m = 1$, $V = 1$, with no inhibition, and the broken lines are calculated with the same values of K_m and V, but $K_{si} = 30$

the corresponding plot of s/v against s, which is a parabola instead of the usual straight line. Provided that K_{si} is much larger than K_m (as is usual), the plot of s/v against s is almost straight at low values of s, and can be used in the usual way to estimate V and K_m.

Substrate inhibition is not usually a significant phenomenon if substrate concentrations are kept at or below their likely physiological values, but it can become important at high substrate concentrations and provides a useful diagnostic tool for distinguishing between possible reaction pathways, as I shall discuss in Section 6.5.

5.12 Chemical modification as a means of identifying essential groups

It is common practice to deduce the nature of groups required for enzymic catalysis from observation that activity is lost when certain residues are chemically modified. Unfortunately, however, the fact that a particular residue is essential for catalytic activity does not certify that it plays any part in the catalytic process; it may, for example, be essential for maintaining the structure of the enzyme. Nonetheless, identification of the essential groups in an enzyme is an important step in characterizing the mechanism, and many reagents are now available for modifying specific types of residue. Tsou (1962) has provided a theoretical framework for analysing chemical modification experiments, but his paper is largely unknown and indeed the logic used for interpreting such experiments is often loose to the point of non-existence.

The simplest case to consider is one in which there are λ groups on each monomeric enzyme molecule that react at the same rate with the modifying agent, and μ of these λ groups are essential for catalytic activity. After modification of an average of γ groups on each molecule, the probability that any particular group has been modified is γ/λ, and the probability that it remains unmodified is $1 - \gamma/\lambda$. For the enzyme molecule to retain activity, *all* of its μ essential groups must be unmodified, for which the probability is $(1 - \gamma/\lambda)^{\mu}$. Thus the fraction α of activity remaining after modification of γ groups per molecule must be

$$\alpha = (1 - \gamma/\lambda)^{\mu}$$

Hence

$$\alpha^{1/\mu} = 1 - \gamma/\lambda$$

and a plot of $\alpha^{1/\mu}$ against γ should be a straight line. One should be able therefore to determine μ by plotting α, $\alpha^{1/2}$, $\alpha^{1/3}$ etc. in turn against γ and choosing the best straight line.

One may object to this treatment that not all of the modification reactions may proceed at the same rate, and they may anyway not be independent, that is, modification of one group may alter the rates at which neighbouring groups are modified. For a full discussion of these and other complications, Tsou's paper should be consulted, but two additional classes of groups can be accommodated without losing the essential simplicity of the method. If there are ξ non-essential groups that react rapidly compared with the essential groups, these will result in an initial region of the plot (regardless of μ) in which γ increases with no decrease in $\alpha^{1/\mu}$. Further groups — whether essential or not — that react slowly compared with the fastest-reacting essential groups will not become appreciably modified until most of the activity has been lost; they must therefore be difficult or impossible to detect. In practice, therefore, a Tsou plot is likely to resemble the one shown in *Figure 5.6*, which was used by Paterson and Knowles (1972) as evidence that at least two carboxyl groups are essential to the activity of pepsin, that three non-essential groups were modified very rapidly in their experiments, and that the two

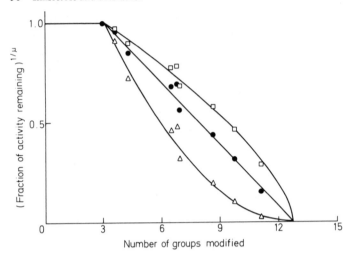

Figure 5.6 Tsou plot for determining the number of essential groups. The plot shows date of Paterson and Knowles (1972) for the inactivation of pepsin by trimethyloxonium fluoroborate, a reagent that reacts specifically with carboxyl groups. The data are plotted for $\mu = 1$ ($\triangle$), $\mu = 2$ ($\bullet$) and $\mu = 3$ ($\square$), and the straight line observed for $\mu = 2$ indicates that at least two carboxyl groups are essential to the activity of pepsin

essential groups formed part of a class of ten that were modified at similar rates.

Pepsin is a monomeric enzyme, but Tsou's analysis can be extended to oligomeric enzymes without difficulty provided that one can assume that the subunits react independently with the modifying agent and that inactivation of one subunit does not affect the activity of the others. With these assumptions, oligomeric enzymes can be treated in the same way as monomeric enzymes, except that μ is now the number of essential groups *per subunit*, even though λ, μ and ξ are still defined *per molecule* (Norris and Brocklehurst, 1976).

Problems

5.1   In the experiments of Kitz and Wilson (1962) the fastest inactivation measured was estimated to have $k_{+2} = 5 \times 10^{-3}$ s^{-1}, $K_i = 0.1$ mM, for the mechanism shown on p. 73. Assuming that K_i can be expressed as $(k_{-1} + k_{+2})/k_{+1}$, and that Kitz and Wilson were wrong to assume a pre-equilibrium (i.e. to assume $K_i = k_{-1}/k_{+1}$), as argued by Childs and Bardsley (1975), estimate the value of k_{+1}. Comment critically on your result and the assumptions that led to it in the light of the fact that second-order rate constants for specific binding of small molecules to proteins have typically been found to be of the order of 10^6 M^{-1} s^{-1} or greater.

5.2 Early work on a racemic substrate for an enzyme reveals that after incubation of the racemic mixture with enzyme, the L enantiomer is completely converted into product whereas the D enantiomer is unchanged. On the basis of this result the kinetics of the reaction are analysed with the assumption that the D enantiomer has no effect on the enzyme and the Michaelis constant for the L enantiomer is estimated to be 2 mM. Subsequent work makes it more reasonable to suppose that the D enantiomer was acting as a competitive inhibitor with K_i equal to the K_m value of the L enantiomer. How should the original estimate of K_m be revised in the light of this information?

5.3 The following data show the initial rates (in arbitrary units) measured for an enzyme-catalysed reaction at various concentrations i and s of inhibitor and substrate respectively. What information can be deduced about the type of inhibition? Comment critically on the design of the experiment.

i (mM)	$s = 1$ mM	$s = 2$ mM	$s = 3$ mM
0	2.36	3.90	5.30
1	1.99	3.35	4.40
2	1.75	2.96	3.98
3	1.60	2.66	3.58
4	1.37	2.35	3.33

5.4 When Norris and Brocklehurst (1976) treated urease with 2,2′-dipyridyl disulphide, a compound that reacts specifically with thiol groups, they observed that the relative catalytic activity α decreased as the number of groups modified per molecule, γ, increased, as detailed in the table below. Assuming that urease has six subunits per molecule that act independently both in the catalytic and in the modification reactions, estimate (a) the number of essential thiol groups per subunit, and (b) the number of inessential thiol groups per subunit that are modified rapidly in comparison with the essential groups.

γ	α	γ	α	γ	α
0.0	1.000	23.0	0.957	27.0	0.547
2.0	1.000	24.0	0.896	27.5	0.442
4.0	1.000	25.0	0.853	28.0	0.353
18.0	1.000	25.5	0.799	29.0	0.198
20.0	1.000	26.0	0.694	29.5	0.104
22.0	0.982	26.5	0.597	30.0	0.011

5.5   Some authors use the symbols K_{is} and K_{ii} for the quantities referred to in this chapter as K_i and K'_i. The second subscripts s and i stand respectively for slope and intercept, and they refer to the fact that if the slopes and intercepts of one of the primary plots described in Section 2.5 are replotted against the inhibitor concentration, they provide values of the two inhibition constants. (a) Which primary plot is referred to? (b) Which intercept (ordinate or abscissa) is replotted? (c) How do the inhibition constants appear in the secondary plot?

5.6   At any given ratio of inhibitor concentration to the appropriate inhibition constant, a competitive inhibitor decreases the rate more than an uncompetitive inhibitor if the substrate concentration is less than K_m ; the reverse is true if the substrate concentration is greater than K_m . Prove this relationship algebraically and explain it conceptually, i.e. without reference to algebra.

5.7   Intestinal aminotripeptidase catalyses the hydrolysis of tripeptides into their N-terminal amino acids and C-terminal dipeptides. Doumeng and Maroux (1979) reported the following kinetic parameters for the hydrolysis of various substrates at pH 7.0 and 37 $^\circ$C:

Substrate	k_{cat} (s^{-1})	K_m (mM)
L-Pro–Gly–Gly	385	1.3
L-Leu–Gly–Gly	190	0.55
L-Ala–Gly–Gly	365	1.4
L-Ala–L-Ala–L-Ala	298	0.52

(a) Which substrate would be expected to be hydrolysed most rapidly in the initial stages of reaction after addition of a sample of the enzyme to a mixture of all four substrates in equimolar concentrations?

(b) When L-Ala–Gly–Gly was studied as a competitive inhibitor of the hydrolysis of L-Pro–Gly–Gly, it was found to have $K_i = 1.4$ mM . Is this value consistent with the view that the enzyme has a single active site at which both substrates are hydrolysed?

Chapter 6

Two-substrate reactions

6.1 Introduction

Much of the earlier part of this book has been concerned with reactions of a single substrate and a single product. Actually, such reactions are rather rare in biochemistry, being confined to a few isomerizations, such as the interconversion of glucose 1-phosphate and glucose 6-phosphate, catalysed by phosphoglucomutase. In spite of this, the development of enzyme kinetics was greatly simplified by two facts: first, that many hydrolytic enzymes can normally be treated as single-substrate enzymes, because the second substrate, water, is always present in such large excess that its concentration can be treated as a constant; second, most enzymes behave much like single-substrate enzymes if only one substrate concentration is varied. This will be clear from the rate equations to be introduced in this chapter, but there is a proviso in this case that K_m for a single substrate has a physical meaning only if the constant conditions are well defined.

There are three principal steady-state kinetic methods for elucidating the order of addition of substrates and release of products: measurement of initial rates in the absence of product; testing the nature of product inhibition; and tracer studies with radioactively labelled substrates. These methods are discussed in this chapter, using a general reaction with two substrates and two products as an example:

$$A + B \rightleftharpoons P + Q$$

This type of reaction is by far the commonest in biochemistry, as it accounts for about 60% of all known enzyme-catalysed reactions. More complex reactions also occur, with as many as four or more substrates, but these reactions can be studied by a simple extension of the principles developed for the study of two-substrate two-product reactions. Even the simple reaction above can take place in a wide variety of ways, but I shall confine discussion to a small number of important cases, rather than attempt an exhaustive treatment. Not

only would such an attempt be self-defeating, because nature can be
relied on to provide examples that fall outside any such 'exhaustive'
treatment; but, more important, the reader who understands the
methods used to discriminate between the simple cases is well equipped
to adapt them to special experimental circumstances and to study the
more detailed discussions that may be found elsewhere.

6.2 Types of mechanism

Almost all two-substrate two-product reactions are formally *group-
transfer* reactions, that is, ones in which a group G is transferred
from one radical, X, to another, Y:

$$GX + Y \rightleftharpoons X + GY$$

This symbolism was introduced by Wong and Hanes (1962) in an
important paper that laid the foundation of the modern classification
of mechanisms. It is convenient for discussing the mechanisms themselves, but becomes rather cumbersome if one attempts to use it
when discussing rate equations. I shall therefore return to the use of
single letters, A, B . . . for substrates, and P, Q . . . for products, later
in this chapter.

Wong and Hanes (1962) showed that most reasonable mechanisms
for group-transfer reactions could be regarded as special cases of a
general mechanism. Perhaps fortunately, enzymes that require the
complete mechanism seem to be very rare, however, and some sup-
posed examples, such as pyruvate carboxylase, may well have been
misinterpreted (*see* Warren and Tipton, 1974). Accordingly, I shall
discuss the three simplest group-transfer mechanisms as separate cases.

The main division is between mechanisms that proceed through a
ternary complex, EGXY, so called because it contains the enzyme
and both substrates in a single species, and those that proceed through
a *substituted enzyme*, EG, which contains the enzyme and the transferred group but neither of the two complete substrates. Early
workers, such as Woolf (1929, 1931) and Haldane (1930), assumed
that the reaction would proceed through a ternary complex, and that
this could be formed by way of either of the two binary complexes
EGX and EY. In other words, the substrates could bind to the enzyme
in *random order*, as illustrated in *Figure 6.1*. The rigorous steady-state
equation for this mechanism is complex, and includes terms in $[GX]^2$
and $[Y]^2$. The contribution of such terms to the rate may well be
very slight, however, and Gulbinsky and Cleland (1968) have shown
by computer simulation that for wide varieties of values assumed for
the rate constants, the experimental rate equation is of the same form
as one derived on the assumption that all steps except the interconversion of EXG·Y and EX·GY are at equilibrium. If this assumption

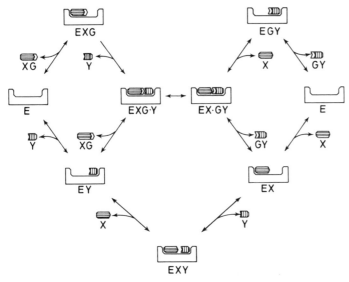

Figure 6.1 Ternary-complex mechanism for a two-substrate two-product reaction, assuming that the substrates bind to and the products are released from the enzyme in random order. The non-productive complex EXY is likely to be kinetically significant only at high concentrations of both X and Y, and is often ignored in simple treatments

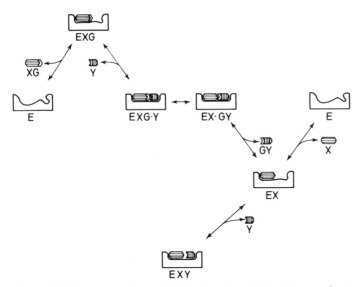

Figure 6.2 Ternary-complex mechanism for a two-substrate two-product reaction, assuming that the substrates bind to and the products are released from the enzyme in a compulsory order. This is presumed to arise because the binding site for the second substrate becomes recognizable only after an appropriate change in conformation has been induced by the binding of the first substrate. The non-productive complex EXY is likely to be kinetically significant only at high concentrations of both X and Y, and is often ignored in simple treatments

is made, no squared terms appear in the rate equation, and for simplicity I shall make the rapid-equilibrium assumption in discussing this mechanism. However, it must be emphasized that the fact that experimental observations often fail to disprove the rapid-equilibrium assumption does *not* imply that the assumption is correct. The step EXG·Y ⇌ EX·GY cannot be detected by steady-state measurements, but it is logical to include it in the random-order mechanism because it is formally treated as rate-determining in deriving the rate equation.

The non-productive complex, EXY, is not a necessary feature of the random-order mechanism, but it can normally be expected to occur, because if both EY and EX are significant intermediates there is no reason to exclude EXY. Another non-productive complex (not included in *Figure 6.1*) is possible if the transferred group G is not too bulky: EXG·GY can result from the binding of GY to EGX or of GX to EGY. This is less likely than the formation of EXY, however.

It is now generally recognized that many enzymes cannot be regarded as rigid templates, as suggested by *Figure 6.1*. Instead, it is likely that the conformations of both enzyme and substrate are altered on binding, in accordance with the 'induced-fit' hypothesis of Koshland (1958, 1959a, b; *see also* Section 8.6). It may well happen therefore that no binding site exists on the enzyme for one of the two substrates until the other is bound. In such cases, there is a *compulsory order* of binding, as illustrated in *Figure 6.2*. If both substrates and products are considered, four different orders are possible, but the induced-fit explanation of compulsory-order mechanisms leads us to expect that the reverse reaction should be structurally analogous to the forward reaction, so that the second product ought to be the structural analogue of the first substrate. Thus only two of the four possibilities are very likely. In NAD-dependent dehydrogenase reactions, for example, the coenzymes are often found to be first substrate and second product. For the same reason as in the random-order case, the non-productive complex EXY is likely to occur in compulsory-order mechanisms.

Early in the development of multiple-substrate kinetics, Doudoroff, Barker and Hassid (1947) showed by isotope-exchange studies that the reaction catalysed by sucrose glucosyltransferase proceeded through a substituted-enzyme intermediate rather than a ternary complex. Since then, studies with numerous and disparate enzymes, including α-chymotrypsin, transaminases and flavoenzymes, have shown that the *substituted-enzyme mechanism*, illustrated in *Figure 6.3*, is common and important. In the ordinary form of this mechanism, occurrence of a ternary complex is structurally impossible because the binding sites for X and Y are either the same or overlapping. For the transaminases, a major group of enzymes that obey this mechanism, all four reactants are structurally similar, so it is

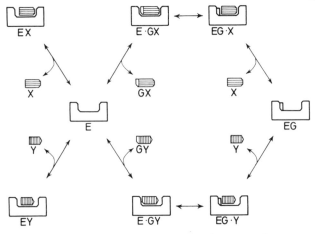

Figure 6.3 Substituted-enzyme mechanism for a two-substrate two-product reaction. The sites for X and Y are assumed to coincide or overlap, in contrast to the arrangement for ternary-complex mechanisms (*Figures 6.1* and *6.2*). The non-productive complexes EX and EY are likely to be kinetically significant only at high concentrations of X and Y, respectively, and are often ignored in simple treatments

reasonable to expect the binding sites for X and Y to be virtually identical and the second half of the reaction to be very similar to the reverse of the first half, e.g.

glutamate + pyridoxal–enzyme   ⇌   intermediates
  ⇌   2-oxoglutarate + pyridoxamine–enzyme

oxaloacetate + pyridoxamine–enzyme   ⇌   intermediates
  ⇌   aspartate + pyridoxal–enzyme

In this mechanism, it is usually possible for substrates to bind to the 'wrong' form of the enzyme, so that there is substrate inhibition at high concentrations (Section 6.5). This is almost always true of E, X and Y, as indicated in *Figure 6.3*, but occurs less often with EG, GX and GY because of steric interference between two G groups.

The substituted-enzyme mechanism shown in *Figure 6.3* is a compulsory-order mechanism, but this is less noteworthy than with ternary-complex mechanisms because there is only one mechanistically reasonable order, and no random-order alternative: even if X and Y bind to E, there is no way for the resulting complexes to break down to give GX or GY. The kinetic properties of the random-order substituted-enzyme mechanism have nonetheless been described, together with those of numerous other mechanisms that are difficult to visualize in chemical terms (Fisher and Hoagland, 1968; Sweeny and Fisher, 1968). The method of King and Altman (1956) can be applied as readily to unreasonable as to reasonable mechanisms, and if one regards kinetics as a branch of algebra, largely unrelated to chemistry, one is liable to be faced with a bewildering array of mechanisms to be considered.

In a substituted-enzyme mechanism the group G is transferred twice, first from the substrate GX to the free enzyme E, then from the substituted enzyme EG to the second substrate Y. For this reason, Koshland (1954) introduced the term *double-displacement* reaction for this type of mechanism. Conversely, ternary-complex mechanisms, in which G is transferred once only, are *single-displacement* reactions. This terminology is still sometimes used, especially in non-kinetic contexts. It leads naturally to consideration of the stereochemistry of group-transfer reactions, which is discussed in detail by Koshland (1954) and is explored in Problem 6.1 at the end of this chapter.

At one time it seemed possible to express experimental results in terms of some broad generalizations, for example that kinases followed random-order ternary-complex mechanisms, NAD-dependent dehydrogenases followed compulsory-order ternary-complex mechanisms (with NAD^+ bound first and NADH released last), and transaminases followed substituted-enzyme mechanisms. This sort of classification is not wholly wrong, but it is now clear that it is oversimplified. For example, alcohol dehydrogenase from horse liver, once regarded as an archetypal example of an enzyme obeying a compulsory-order ternary-complex mechanism, is now thought to bind its substrates in random order but to release the products in a compulsory order (Hanes *et al.*, 1972). In the strictest sense compulsory-order ternary-complex mechanisms may not occur at all, but they remain useful as a basis for discussion.

6.3 Rate equations

Steady-state kinetic measurements have proved to be of enormous value in distinguishing between the various reaction mechanisms for group-transfer reactions. The development of these methods was a considerable task, on account of the large number of possibilities and the relatively small kinetic differences between them. Segal, Kachmar and Boyer (1952) were among the first to recognize the need for a systematic approach, and derived the equations for several mechanisms. Subsequently, Alberty (1953, 1958) and Dalziel (1957) made major advances in the understanding of group-transfer reactions, and introduced most of the methods described in this chapter.

As all steady-state methods for distinguishing between mechanisms depend on differences between the complete rate equations, it is appropriate to give a brief account of these equations before discussing methods. The equation for the compulsory-order ternary-complex mechanism was derived in Section 4.1 as an illustration of the King–Altman method, and had the form

$$v = \frac{e_0(c_1 ab - c_2 pq)}{c_3 + c_4 a + c_5 b + c_6 p + c_7 q + c_8 ab + c_9 ap + c_{10} bq + c_{11} pq + c_{12} abp + c_{13} bpq} \tag{6.1}$$

This equation contains 13 coefficients, but these were defined in terms of only eight rate constants; there must therefore be relationships between the coefficients that are not explicit in the equation. Moreover, the coefficients are without obvious mechanistic meaning. Numerous systems have been used for rewriting rate equations in more meaningful terms (*see*, for example, Alberty, 1953; Dalziel, 1957; Bloomfield, Peller and Alberty, 1962; Cleland, 1963; Mahler and Cordes, 1966). Of these the simplest to understand and use is probably that of Cleland, modified slightly in this book to accord with the Recommendations of the Commission on Biochemical Nomenclature (1973). For any mechanism, maximum velocities in the forward and reverse directions are written as V^f and V^r, respectively, though the superscripts can be omitted in unambiguous cases; for each substrate there is a *Michaelis constant*, K_m^A, K_m^B, etc. and an *inhibition constant* K_i^A, K_i^B, etc. The meanings of these will become clear in later sections of this chapter, but in general the Michaelis constants correspond to K_m in a one-substrate reaction, and the inhibition constants are related to (but not necessarily equal to) the K_i and K_i values obtained when the particular substrate is used as a product inhibitor of the reverse reaction. Under some circumstances the inhibition constants are true substrate-dissociation constants, and they are sometimes written therefore as K_s^A, etc., rather than K_i^A, etc.

With this system, equation 6.1 becomes

$$v = \frac{\dfrac{V^f ab}{K_i^A K_m^B} - \dfrac{V^r pq}{K_m^P K_i^Q}}{1 + \dfrac{a}{K_i^A} + \dfrac{K_m^A b}{K_i^A K_m^B} + \dfrac{K_m^Q p}{K_m^P K_i^Q} + \dfrac{q}{K_i^Q} + \dfrac{ab}{K_i^A K_m^B} + \dfrac{K_m^Q ap}{K_i^A K_m^P K_i^Q}} \tag{6.2}$$

$$+ \frac{K_m^A bq}{K_i^A K_m^B K_i^Q} + \frac{pq}{K_m^P K_i^Q} + \frac{abp}{K_i^A K_m^B K_i^P} + \frac{bpq}{K_i^B K_m^P K_i^Q}$$

where the kinetic parameters have the values shown in *Table 6.1*. Although this equation has a complex appearance, it contains more regularities than may be immediately apparent: first, the terms that contain q are in general similar to those that contain a, whereas the terms that contain p are similar to those that contain b; second,

TABLE 6.1 Definitions of kinetic parameters for compulsory-order mechanisms

The table shows the relationships between the parameters that appear in equations 6.2 and 6.4 and the rate constants for the individual steps of the two principal compulsory-order mechanisms for group-transfer reactions. Although equation 6.4 does not contain K_i^B it can be rewritten so that it does by means of the identity $K_i^A K_m^B / K_i^P K_m^Q = K_m^A K_i^B / K_m^P K_i^Q$.

	Ternary-complex mechanism	*Substituted-enzyme mechanism*
	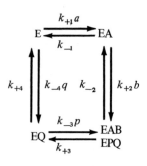	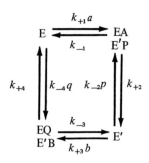
V^f	$\dfrac{k_{+3} k_{+4} e_0}{k_{+3} + k_{+4}}$	$\dfrac{k_{+2} k_{+4} e_0}{k_{+2} + k_{+4}}$
V^r	$\dfrac{k_{-1} k_{-2} e_0}{k_{-1} + k_{-2}}$	$\dfrac{k_{-1} k_{-3} e_0}{k_{-1} + k_{-3}}$
K_m^A	$\dfrac{k_{+3} k_{+4}}{k_{+1}(k_{+3} + k_{+4})}$	$\dfrac{(k_{-1} + k_{+2})k_{+4}}{k_{+1}(k_{+2} + k_{+4})}$
K_m^B	$\dfrac{(k_{-2} + k_{+3})k_{+4}}{k_{+2}(k_{+3} + k_{+4})}$	$\dfrac{k_{+2}(k_{-3} + k_{+4})}{(k_{+2} + k_{+4})k_{+3}}$
K_m^P	$\dfrac{k_{-1}(k_{-2} + k_{+3})}{(k_{-1} + k_{-2})k_{-3}}$	$\dfrac{(k_{-1} + k_{+2})k_{-3}}{(k_{-1} + k_{-3})k_{-2}}$
K_m^Q	$\dfrac{k_{-1} k_{-2}}{(k_{-1} + k_{-2})k_{-4}}$	$\dfrac{k_{-1}(k_{-3} + k_{+4})}{(k_{-1} + k_{-3})k_{-4}}$
K_i^A	k_{-1}/k_{+1}	k_{-1}/k_{+1}
K_i^B	$(k_{-1} + k_{-2})/k_{+2}$	k_{-3}/k_{+3}
K_i^P	$(k_{+3} + k_{+4})/k_{-3}$	k_{+2}/k_{-2}
K_i^Q	k_{+4}/k_{-4}	k_{+4}/k_{-4}

the reactants that appear in the denominator of any term as superscripts also appear in the numerator of the same term, either as
concentrations or as superscripts.

The corresponding equation for the random-order ternary-complex
mechanism is

$$v = \frac{\dfrac{V^f ab}{K_i^A K_m^B} - \dfrac{V^r pq}{K_m^P K_i^Q}}{1 + \dfrac{a}{K_i^A} + \dfrac{b}{K_i^B} + \dfrac{p}{K_i^P} + \dfrac{q}{K_i^Q} + \dfrac{ab}{K_i^A K_m^B} + \dfrac{pq}{K_m^P K_i^Q}} \qquad (6.3)$$

It is perhaps surprising that the simpler equation should refer to the
more complicated mechanism; the explanation is that equation 6.3,
unlike equation 6.2, is derived by assuming that all steps apart from
the interconversion of the ternary complexes EAB and EPQ are at
equilibrium. With this assumption, K_i^A, K_i^B, K_i^P and K_i^Q are the
dissociation constants of EA, EB, EP and EQ, respectively; K_m^A and
K_m^B are the dissociation constants of EAB for loss of A and B, respectively; K_m^P and K_m^Q are the dissociation constants of EPQ for loss of
P and Q, respectively. (Although K_m^A and K_m^Q do not appear explicitly
in equation 6.3, they can be introduced because in this mechanism
$K_i^A K_i^B$ is interchangeable with $K_i^A K_m^B$, and $K_i^P K_i^Q$ with $K_m^P K_i^Q$. These
substitutions cannot be made in equation 6.2, because the compulsory-
order mechanism is not symmetrical in A and B or in P and Q, and this
provides a further reason for the greater complexity of equation
6.2.) Equation 6.3 may well apply within experimental error whether
the equilibrium assumption is correct or not, however, and the
Michaelis and inhibition constants cannot therefore be interpreted in
general as true dissociation constants.

The equation for the substituted-enzyme mechanism is

$$v = \frac{\dfrac{V^f ab}{K_i^A K_m^B} - \dfrac{V^r pq}{K_i^P K_m^Q}}{\dfrac{a}{K_i^A} + \dfrac{K_m^A b}{K_i^A K_m^B} + \dfrac{p}{K_i^P} + \dfrac{K_m^P q}{K_i^P K_m^Q} + \dfrac{ab}{K_i^A K_m^B} + \dfrac{ap}{K_i^A K_i^P} + \dfrac{K_m^A bq}{K_i^A K_m^B K_i^Q} + \dfrac{pq}{K_i^P K_m^Q}} \qquad (6.4)$$

where the kinetic parameters are again defined in *Table 6.1*. In coefficient form, this equation is the same as equation 6.1 without the
constant and the terms in *abp* and *bpq* in the denominator, but the
relationships between the parameters are different, and equation 6.4
has $K_i^P K_m^Q$ wherever $K_m^P K_i^Q$ might be expected by analogy with
equation 6.2.

Although *Table 6.1* provides the values of the kinetic parameters
in terms of rate constants, it does not give the reverse relationships.

One may wonder, therefore, whether it is possible to calculate the individual rate constants from measurements of maximum velocities, Michaelis constants and inhibition constants. Provided that the enzyme concentration e_0 (expressed in mol l^{-1}) is known, this is in fact possible for the compulsory-order ternary-complex mechanism, but not for the substituted-enzyme mechanism (Cleland, 1963). The relevant relationships for the former mechanism are shown in *Table 6.2*, but they should be used with caution, because they make no

TABLE 6.2 Calculation of rate constants from kinetic parameters

The table is obtained by rearranging the definitions given in *Table 6.1* for the kinetic parameters of the compulsory-order ternary-complex mechanism. Analogous rearrangement of the definitions for the substituted-enzyme mechanism is not possible.

Rate constant	Expression
k_{+1}	$V^{\mathrm{f}}/K_{\mathrm{m}}^{\mathrm{A}} e_0$
k_{-1}	$V^{\mathrm{f}} K_{\mathrm{i}}^{\mathrm{A}}/K_{\mathrm{m}}^{\mathrm{A}} e_0$
k_{+2}	$V^{\mathrm{f}}(k_{-2} + k_{+3})/k_{+3} K_{\mathrm{m}}^{\mathrm{B}} e_0$
k_{-2}	$V^{\mathrm{f}} V^{\mathrm{r}} K_{\mathrm{i}}^{\mathrm{A}}/e_0 (V^{\mathrm{f}} K_{\mathrm{i}}^{\mathrm{A}} - V^{\mathrm{r}} K_{\mathrm{m}}^{\mathrm{A}})$
k_{+3}	$V^{\mathrm{f}} V^{\mathrm{r}} K_{\mathrm{i}}^{\mathrm{Q}}/e_0 (V^{\mathrm{r}} K_{\mathrm{i}}^{\mathrm{Q}} - V^{\mathrm{f}} K_{\mathrm{m}}^{\mathrm{Q}})$
k_{-3}	$V^{\mathrm{r}}(k_{-2} + k_{+3})/k_{-2} K_{\mathrm{m}}^{\mathrm{P}} e_0$
k_{+4}	$V^{\mathrm{r}} K_{\mathrm{i}}^{\mathrm{Q}}/K_{\mathrm{m}}^{\mathrm{Q}} e_0$
k_{-4}	$V^{\mathrm{r}}/K_{\mathrm{m}}^{\mathrm{Q}} e_0$

allowance for the possibility that there may be more than a minimum number of steps in the mechanism: if any of the binary and ternary complexes isomerize, the form of the steady-state rate equation is unaffected, but the interpretation of the parameters is different and some or all of the relationships in *Table 6.2* become invalid. For the random-order ternary-complex mechanism, it is obvious that none of the rate constants apart from those for the rate-limiting inter-conversion of ternary complexes can be determined from steady-state measurements if the rapid-equilibrium assumption is correct. If this assumption is incorrect and there are detectable deviations from equation 6.3, it may be possible to use curve-fitting techniques to deduce information about the rate constants (Cornish-Bowden and Wong, 1978).

6.4 Initial-velocity measurements in the absence of products

If no products are included in the reaction mixture, the initial velocity for a reaction following the compulsory-order ternary-complex mechanism is given by the equation

$$v = \frac{Vab}{K_i^A K_m^B + K_m^B a + K_m^A b + ab} \tag{6.5}$$

This equation is derived from equation 6.2 by omitting terms that contain p or q, writing V^f as V as there is no ambiguity in this case, and multiplying all terms in both numerator and denominator by $K_i^A K_m^B$, to remove the fractions. The meanings of V, K_i^A, K_m^A and K_m^B become apparent if the equation is examined at extreme values of a and b. If both a and b are very large, all terms that do not contain both of them become negligible, and the equation simplifies to $v = V$, so V has a meaning precisely analogous to its meaning in a single-substrate reaction: it is the limiting rate when *both* substrates are saturating. Similarly, we can determine the meanings of the Michaelis constants by considering the effect of making one substrate concentration large while keeping the other moderate. If b is very large, terms that do not contain it can be neglected, and equation 6.5 simplifies to the Michaelis–Menten equation, with K_m^A as the Michaelis constant, i.e.

$$v = \frac{Va}{K_m^A + a}$$

Thus K_m^A may be defined as the limiting Michaelis constant for A when B is saturating. Similarly, K_m^B is the limiting Michaelis constant for B when A is saturating. K_i^A is *not* the same as K_m^A, and its meaning can be seen by considering the effect on equation 6.5 of making b very small (but not zero), so that terms in the denominator that contain b may be neglected. Then

$$v = \frac{(Vb/K_m^B)a}{K_i^A + a}$$

So K_i^A is the limiting value of the Michaelis constant for A when b approaches zero. It is also the true equilibrium dissociation constant of EA, because when b approaches zero the rate of reaction of B with EA must also approach zero; the binding of A to E can then be maintained at equilibrium and the Michaelis–Menten assumption of equilibrium binding is valid in this instance. K_i^B does not appear in equation 6.5, because B does not bind to the free enzyme. It does, however, occur in the equation for the complete reversible reaction, equation 6.2, and its magnitude affects the behaviour of B as an inhibitor of the reverse reaction. Although equation 6.5 is not symmetrical in A and B, because $K_i^A K_m^B$ is not the same as $K_m^A K_i^B$, it is symmetrical in *form*; measurement of initial rates in the absence of products does not therefore distinguish A from B.

If the concentration of one substrate is varied at constant (but not necessarily very high or very low) concentrations of the other, equation 6.5 still has the form of the Michaelis–Menten equation with respect to the varied substrate. For example, if a is varied at constant b, terms that do not contain a are constant, and equation 6.5 can be rearranged into the form

$$ v = \frac{\left(\dfrac{Vb}{K_m^B + b}\right) a}{\left(\dfrac{K_i^A K_m^B + K_m^A b}{K_m^B + b}\right) + a} = \frac{V^{app} a}{K_m^{app} + a} $$

V^{app} and K_m^{app}, the apparent values of V and K_m, are functions of b:

$$ V^{app} = \frac{Vb}{K_m^B + b} \tag{6.6} $$

$$ K_m^{app} = \frac{K_i^A K_m^B + K_m^A b}{K_m^B + b} \tag{6.7} $$

$$ V^{app}/K_m^{app} = \frac{(V/K_m^A)b}{(K_i^A K_m^B /K_m^A) + b} \tag{6.8} $$

In a typical experiment, various values of b would be used, and at each of these the rate would be determined at various values of a. Then V^{app} and K_m^{app} can be determined at each b value by one of the methods discussed in Section 2.5, for example by a plot of a/v against a. Such a plot is called a *primary plot*, to distinguish it from the secondary plots that will be described shortly. *Figure 6.4* shows a typical set of primary plots for an enzyme that obeys equation 6.5. The point of intersection of the lines must occur to the left of the a/v axis (contrast the substituted-enzyme mechanism, *Figure 6.6*, below), as $a = -K_i^A$, $a/v = (K_m^A - K_i^A)/V$ at this point. Whether it occurs above or below the a axis depends on the relative magnitudes of K_i^A and K_m^A.

Equations 6.6 and 6.8 are of the same form as the Michaelis–Menten equation, that is, plots of V^{app} or V^{app}/K_m^{app} against b

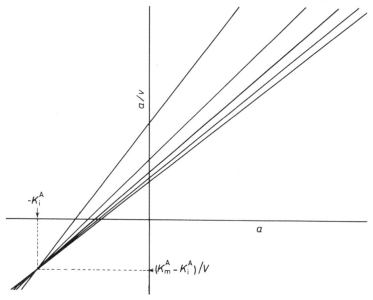

Figure 6.4 Primary plots of a/v against a at various values of b, for ternary-complex mechanisms, ignoring substrate inhibition. Plots of b/v against b at various values of a are similar

describe rectangular hyperbolas through the origin, and they can be analysed in exactly the same way. Thus equation 6.6 can be written as

$$\frac{b}{V^{app}} = \frac{K_m^B}{V} + \frac{1}{V} \cdot b$$

so that a *secondary plot* of b/V^{app} against b is a straight line of slope $1/V$ and intercept K_m^B/V on the b/V^{app} axis. Similarly, equation 6.8 gives

$$\frac{b K_m^{app}}{V^{app}} = \frac{K_i^A K_m^B}{V} + \frac{K_m^A}{V} \cdot b$$

so that a secondary plot of $b K_m^{app}/V^{app}$ against b is a straight line of slope K_m^A/V and intercept $K_i^A K_m^B/V$ on the $b K_m^{app}/V^{app}$ axis. All four parameters, V, K_i^A, K_m^A and K_m^B, can readily be calculated from these plots, which are illustrated in *Figure 6.5*.

Equation 6.7 also describes a rectangular hyperbola, but the curve does not pass through the origin. Instead, K_m^{app} approaches K_i^A as b approaches zero. It is thus a three-parameter hyperbola, and cannot be redrawn as a straight line. As in other cases, K_m^{app} is a less convenient parameter to examine than K_m^{app}/V^{app}.

One can equally well treat B as the variable substrate instead of A, making primary plots of b/v against b at the different values of a.

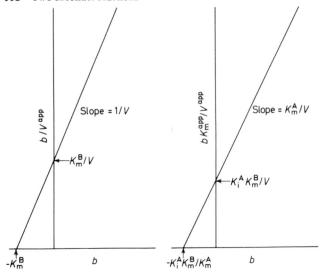

Figure 6.5   Secondary plots for ternary-complex mechanisms. The plot of b/V^{app} against b is also applicable to substituted-enzyme mechanisms

The analysis is the same, and so there is no need to describe it again. The only important difference is that K_i^B does not occur in equation 6.5, and $K_i^A K_m^B / K_m^A$ occurs wherever one might expect K_i^B from simple interchange of A and B.

For the random-order ternary-complex mechanism, the complete rate equation, equation 6.3, again simplifies to equation 6.5 if terms in p and q are omitted. It is, therefore, impossible to tell whether the order of binding of substrates is compulsory or random from measurements of the initial rate in absence of products. Unlike the compulsory-order mechanism, however, the random-order mechanism is symmetrical in A and B, and $K_i^A K_m^B$ is identical to $K_m^A K_i^B$. If the equilibrium assumption is correct, that is, if the breakdown of EAB to products is rate-limiting, then K_i^A and K_i^B are the dissociation constants of EA and EB respectively, and K_m^A and K_m^B are the two dissociation constants of EAB, for loss of A and B, respectively. The graphical analysis is the same as for the compulsory-order mechanism.

For the substituted-enzyme mechanism, the initial rate in the absence of products is given by

$$v = \frac{Vab}{K_m^B a + K_m^A b + ab} \tag{6.9}$$

The most striking feature of this equation is the absence of a constant term from the denominator. (Problem 4.3 at the end of Chapter 4 explores the question of why this should be so.) It causes behaviour recognizably different from that seen with ternary-complex mechanisms when either substrate concentration is varied: for example, if a

is varied at a constant value of b, the apparent values of V and K_m are given by

$$V^{app} = Vb/(K_m^B + b)$$

$$K_m^{app} = K_m^A b/(K_m^B + b)$$

$$V^{app}/K_m^{app} = V/K_m^A$$

Only V^{app} behaves in the same way as in ternary-complex mechanisms. The important characteristic is that V^{app}/K_m^{app} is independent of b, with a constant value of V/K_m^A. It is also constant if b is varied at constant a; its value is then V/K_m^B. Primary plots of a/v against a or of b/v against b form a series of straight lines intersecting on the a/v

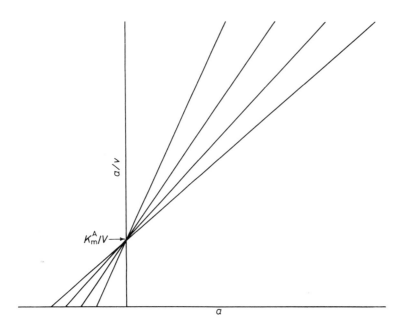

Figure 6.6   Primary plots of a/v against a at various values of b for substituted-enzyme mechanisms, ignoring substrate inhibition. Plots of b/v against b at various values of a are similar

or b/v axis, as shown in *Figure 6.6*. This pattern is readily distinguishable from the pattern of primary plots exhibited by the ternary-complex mechanisms (*Figure 6.4*) unless K_i^A is much smaller than K_m^A.

The only secondary plot required for the substituted-enzyme mechanism is that of b/V^{app} against b, which has the same slope and intercepts as the corresponding plot for the ternary-complex mechanisms (*Figure 6.5*).

6.5 Substrate inhibition

The results in the previous section are strictly valid only at low substrate concentrations because, in all reasonable mechanisms, at least one of the four reactants can bind to the wrong enzyme species. In the substituted-enzyme mechanism, the substrate and product that lack the transferred group (Y and X in the symbolism used in Section 6.2) can be expected to bind to the wrong form of the free enzyme; in the random-order ternary-complex mechanism, the same pair may bind to the wrong binary complexes; and in the compulsory-order ternary-complex mechanism, either the second substrate or the first product may bind to the wrong binary complex. In this last case, substrate inhibition can occur in either the forward or the reverse reaction, but not both, because only one of the two binary complexes is available. For convenience, I shall take B as the reactant that displays substrate inhibition for each mechanism, but the results can readily be transformed for other reactants if required.

The non-productive complex EBQ in the compulsory-order ternary-complex mechanism was considered in Section 4.4. It can be allowed for in the rate equation by multiplying every term in the denominator that refers to EQ by $(1 + k_{+5}b/k_{-5})$, where k_{-5}/k_{+5} is the dissociation constant of EBQ. Equation 6.5 then becomes

$$v = \frac{Vab}{K_i^A K_m^B + K_m^B a + K_m^A b + ab(1 + b/K_{si}^B)} \tag{6.10}$$

where K_{si}^B is a constant that defines the strength of the inhibition. It is *not* the same as k_{-5}/k_{+5}, because the term in ab refers not only to EQ but also to the ternary complex (EAB + EPQ) (*see* the derivation of equation 6.1 in Section 4.1). Depending on the relative amounts of these two complexes in the steady state, K_{si}^B may approximate to k_{-5}/k_{+5}, or it may be much greater. Thus substrate inhibition may not be detectable with this mechanism at any attainable concentration of B.

Substrate inhibition according to equation 6.10 is effective only at high concentrations of A, and thus resembles uncompetitive inhibition. Primary plots of b/v against b are parabolic, with a common intersection point at $b = -K_i^A K_m^B / K_m^A$. Primary plots of a/v against a are linear, but have no common intersection point. These plots are illustrated in *Figure 6.7*.

In the random-order ternary-complex mechanism, the concentration of EQ is zero in the absence of added Q if the rapid-equilibrium assumption holds. As B cannot bind to a species that is absent, substrate inhibition does not occur with this mechanism unless Q is added. If the rapid-equilibrium assumption does not hold, there is no reason why substrate inhibition should not occur, but its nature is difficult to predict with certainty because of the complexity of the rate equation. In this mechanism, EBQ is *not* a dead-end complex because it can be formed from either EB or EQ; it need not therefore be in equilibrium with either.

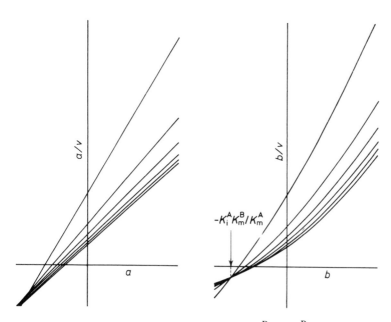

Figure 6.7   Effect of substrate inhibition by B (with $K_{si}^B = 10K_m^B$) on primary plots for ternary-complex mechanisms (cf. *Figure 6.4*)

In the substituted-enzyme mechanism, the non-productive complex EB results from the binding of B to E (B = Y in the symbolism of Section 6.2). It is a dead-end complex, and so it can be allowed for by multiplying terms in the denominator of the rate equation that refer to E by $(1 + b/K_{si}^B)$, where K_{si}^B is the dissociation constant of EB. Equation 6.9 therefore becomes

$$v = \frac{Vab}{K_m^B a + K_m^A b(1 + b/K_{si}^B) + ab} \tag{6.11}$$

Inhibition according to this equation is most effective when a is small, and thus resembles competitive inhibition. Primary plots of b/v against b are parabolic and intersect at a common point on the b/v axis, i.e. at $b = 0$. Primary plots of a/v against a are linear, with no common intersection point, but every pair of lines intersects to the right of the a/v axis, i.e. at a positive value of a. These plots are illustrated in *Figure 6.8*.

Substrate inhibition might seem at first sight to be a tiresome complication in the analysis of kinetic data. Actually, it is very informative, because it accentuates the difference in behaviour predicted for ternary-complex and substituted-enzyme mechanisms, and is usually straightforward to interpret. As substrates normally bind more tightly to the correct enzyme species than to the wrong one, substrate inhibition is rarely severe enough to interfere with the analysis described

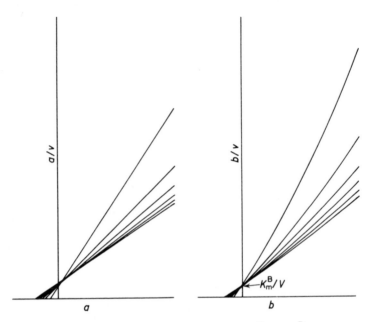

Figure 6.8   Effect of substrate inhibition by B (with $K_{si}^B = 10K_m^B$) on primary plots for substituted-enzyme mechanisms (cf. *Figure 6.6*)

in Section 6.4. Substrate inhibition at low concentrations of the constant substrate provides strong positive evidence for the substituted-enzyme mechanism. In contrast, the observation that primary plots intersect on the a/v or b/v axis is only negative evidence, as it can occur as a special case of a ternary-complex mechanism. When substrate inhibition occurs in a compulsory-order ternary-complex mechanism, it allows the substrate that binds second to be identified, which would otherwise require product-inhibition studies.

6.6 Product inhibition

Product-inhibition studies are among the most useful of methods for elucidating the order of binding of substrates and release of products, as they are both informative and simple to understand. Provided that only one product is added to a reaction mixture, the term in the numerator of the rate equation that refers to the reverse reaction must be zero (except for one-product reactions, which are rare). The only effect of adding product, therefore, is to increase the denominator of the rate equation, that is, to inhibit the forward reaction. The question of whether a product acts as a competitive, uncompetitive or mixed inhibitor cannot be answered in absolute terms, because the answer depends on which substrate is considered to be variable. Once this has been decided, however, the question is straightforward: the denominator of any rate equation can be separated into *variable* and *constant* terms according to whether they contain the variable-substrate concentration or not; the expression for V^{app} depends on the variable terms, whereas the expression for V^{app}/K_m^{app} depends on the constant terms, as in Section 6.4. Now the various kinds of inhibition are classified according to whether they affect V^{app}/K_m^{app} (competitive inhibition), V^{app} (uncompetitive inhibition) or both (mixed inhibition; *see Table 5.1*, p. 79); so a product is a competitive inhibitor if its concentration appears only in constant terms, an uncompetitive inhibitor if it appears only in variable terms and a mixed inhibitor if it appears in both. If the product can combine with only one form of the enzyme, only linear terms in its concentration are possible, and so the inhibition is linear, but non-linear inhibition becomes possible if the product can also bind to 'wrong' enzyme forms to give dead-end complexes.

Let us apply these principles to equation 6.2, the equation for the compulsory-order ternary-complex mechanism, under conditions where P is added to the reaction mixture but Q is not, so that terms containing q can be neglected. If A is the variable substrate, the constant part of the denominator is

$$1 + \frac{K_m^A b}{K_i^A K_m^B} + \frac{K_m^Q p}{K_m^P K_i^Q}$$

and the variable part is

$$\frac{a}{K_i^A}\left(1 + \frac{b}{K_m^B} + \frac{K_m^Q p}{K_m^P K_i^Q} + \frac{bp}{K_m^B K_i^P}\right)$$

Both of these expressions contain p, and therefore P is a mixed inhibitor when A is the variable substrate. By a similar analysis one can show that both P and Q behave as mixed inhibitors when B is the variable substrate. When one considers inhibition by Q with A as variable substrate, however, the results are different: the denominator of equation 6.2 contains no terms in which a and q are multiplied together, though q does occur in terms that do not contain a; thus q

occurs in constant terms only in this case and Q is competitive with respect to A. These results, together with the corresponding ones for the substituted-enzyme mechanism, are summarized in *Table 6.3*. The types of inhibition expected for the random-order ternary-complex mechanism are considered in Problem 6.6 at the end of this chapter.

TABLE 6.3   Product inhibition in the two principal compulsory-order mechanisms

The table shows the type of inhibition expected for each combination of product and variable substrate. The descriptions in parentheses show how the type of inhibition is modified at saturating concentrations of the constant substrate.

Product	Variable substrate	Type of inhibition[†]	
		Ternary-complex mechanism	Substituted-enzyme mechanism
P	A	Mixed (unc.)	Mixed (no inh.)
P	B	Mixed (mixed)	Comp. (comp.)
Q	A	Comp. (comp.)	Comp. (comp.)
Q	B	Mixed (no inh.)	Mixed (no inh.)

[†]The following abbreviations are used: comp., competitive; unc., uncompetitive; no inh., no inhibition

At very high concentrations of the constant substrate, the types of product inhibition become modified, because terms in the rate equation that do not contain the constant-substrate concentration become negligible. For example, if B approaches saturation, the constant part of the denominator of equation 6.2 becomes effectively independent of p, but the variable part does not. Consequently, the mixed inhibition by P with A as variable substrate in the compulsory-order ternary-complex mechanism becomes simple uncompetitive inhibition as B approaches saturation. This result, as well as the corresponding ones for other combinations, is included in *Table 6.3*.

It is a simple matter to predict the product-inhibition characteristics of any mechanism. The most reliable method is to study the form of the complete rate equation, but one can usually arrive at the same conclusions by inspecting the mechanism in the light of the method of King and Altman, as described in Section 4.4. For any combination of product and variable substrate, one must search for a King–Altman pattern that gives rise to a term containing the product concentration but not the variable-substrate concentration; if one is successful the product must appear in the constant part of the denominator. One must then search for a King–Altman pattern that gives rise to a term containing both product and variable-substrate concentrations: if the search is successful the product must appear in the variable part of the denominator. With this information it is a simple matter to use the approach described above to decide on the type of inhibition. In searching for suitable King–Altman patterns, one must remember that product-release steps are irreversible if the product in question is not present in the reaction mixture.

In two-product reactions, uncompetitive inhibition is largely confined to the case mentioned above, inhibition by the first product in a compulsory-order ternary-complex mechanism when the second substrate is saturating. It becomes more common in reactions with three or more products, and occurs with at least one product in all compulsory-order mechanisms for such reactions (p. 127).

6.7 Design of experiments

The design of an experiment to study an uninhibited two-substrate reaction rests on principles similar to those discussed in Section 5.9 for studying simple inhibition. The values of the Michaelis and inhibition constants for the various substrates will not of course be known in advance, and some trial experiments must be done in ignorance. However, a few experiments in which one substrate concentration is varied at each of two concentrations of the other, one as high and the other as low as practically convenient, should reveal the likely range of apparent K_m values for the first substrate. This range can then be used to select the concentrations of this substrate to be used in a more thorough study. The concentrations of the other substrate can be selected similarly on the basis of a converse trial experiment. At each concentration of constant substrate the variable-substrate concentrations should extend from about $0.5K_m^{app}$ to about $10K_m^{app}$ or as high as conveniently possible, as in the imaginary inhibition experiment outlined in *Table 5.3* (p. 90). It is not necessary to have exactly the same set of concentrations of variable substrate at each concentration of constant substrate. It is, however, useful to have sets based loosely on a grid (as in *Table 5.3*), because this allows the same experiment to be plotted both ways, with each substrate designated 'variable' in turn. Note that the labels 'variable' and 'constant' are experimentally arbitrary, and are convenient only for analysing results and for defining what we mean by 'competitive', 'uncompetitive', etc. in reactions with more than one substrate.

The design of product-inhibition experiments for multiple-substrate experiments requires no special discussion beyond that given in Section 5.9 for simple inhibition studies. It should be sufficient to emphasize that the experiment should be carried out in such a way as to reveal whether significant competitive and uncompetitive components are present.

6.8 Isotope exchange

Study of the initial rates of multiple-substrate reactions in both forward and reverse directions, and in the presence and absence of products, will usually eliminate many possible reaction pathways and give a good indication of the gross features of the mechanism, but it will not usually reveal the existence of any minor alternative pathways if these contribute negligibly to the total rate. Further information is therefore required to provide a definitive picture. Even if a clear mechanism does emerge from initial-rate and product-inhibition experiments, it is valuable to be able to confirm its validity independently. The important technique of isotope exchange, which was introduced to enzyme kinetics by Boyer (1959), can often satisfy these requirements.

To apply the results of isotope-exchange experiments, one must normally make two important assumptions. These are usually true and are often merely implied, but it is as well to state them clearly to avoid misunderstanding. The first assumption is that a reaction that involves radioactive substrates follows the same mechanism as the normal reaction, with the same rate constants. In other words, isotope effects are assumed to be negligible. This assumption is usually true, provided that tritium is not used as a radioactive label. Even then, isotope effects are likely to be negligible if the tritium atom is not directly involved in the reaction or in binding the substrate to the enzyme[†]. The second assumption is that the concentrations of all radioactive species are so low that they have no perceptible effect on the concentrations of unlabelled species. This assumption can usually be made to be true, and is important, because it allows labelled species to be ignored in calculating the concentrations of unlabelled species and thus simplifies the analysis considerably.

Isotope exchange can most readily be understood in relation to an example, such as the transfer of a radioactive atom (represented by

---

[†] Isotope effects may also be studied for their own sake, as they can provide valuable information about the breaking or stretching of bonds in the rate-limiting step of a reaction. This is, however, a technique quite distinct from the study of isotope exchange that is discussed in this section, though it is one of great importance in chemistry, and increasing importance in enzymology. For a discussion of hydrogen isotope effects in enzymology *see* Northrop (1977) and other articles in the same volume, and for an illustration of the power of isotope effects in the elucidation of enzyme mechanisms, *see* Albery and Knowles (1976), who used them to obtain a wealth of information about the mechanism of action of triose phosphate isomerase.

an asterisk) from A* to P* in the compulsory-order ternary-complex mechanism:

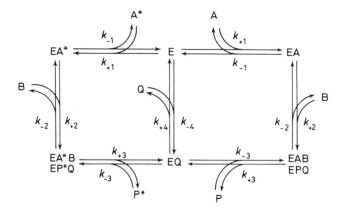

As this exchange requires A* to bind to E, it can occur only if there is a significant concentration of E. The exchange reaction must there-fore be inhibited by high concentrations of either A or Q, as they compete with A* for E. The effects of B and P are more subtle: on the one hand, the exchange reaction includes the binding of B to EA*, and so a finite concentration of B is required. On the other hand, if B and P are present at very high concentrations, the enzyme will exist largely as ternary complex, (EAB + EPQ), and so there will be no E for A* to bind to. One would therefore expect high concentra-tions of B and P to inhibit the exchange and it is not difficult to show that this expectation is correct. The rates of change of labelled inter-mediate concentrations can be written in the usual way, and set to zero according to the steady-state assumption:

$$\frac{d}{dt}[EA^*] = k_{+1}a^*[E] - (k_{-1} + k_{+2}b)[EA^*] + k_{-2}[EA^*B] = 0$$

$$\frac{d}{dt}[EA^*B] = k_{+2}b[EA^*] - (k_{-2} + k_{+3})[EA^*B] + k_{-3}p^*[EQ] = 0$$

These are a pair of simultaneous equations in [EA*] and [EA*B]. The solution for [EA*B], with p^* set to zero, is

$$[EA^*B] = \frac{k_{+1}k_{+2}a^*b[E]}{k_{-1}(k_{-2} + k_{+3}) + k_{+2}k_{+3}b}$$

The initial rate of exchange, v^*, is given by $k_{+3}[EA^*B]$, or

$$v^* = \frac{k_{+1}k_{+2}k_{+3}a^*b[E]}{k_{-1}(k_{-2} + k_{+3}) + k_{+2}k_{+3}b} \tag{6.12}$$

An expression for [E] is required before this equation can be used. As indicated above, the treatment is simplified by assuming that the concentrations of unlabelled species are unaffected by the presence of trace amounts of labelled species; accordingly, the value of [E] is the same as if there were no label. If the experiment is done while the unlabelled reaction is in a steady state, the expression for [E] derived in Section 4.1 must be used. This is unnecessarily complicated, however, because it is usual to study isotope exchange while the unlabelled reaction is at equilibrium. In this case, [E] is given by the expression

$$[E] \; = \; \frac{e_0}{1 + \dfrac{k_{+1}a}{k_{-1}} + \dfrac{k_{+1}k_{+2}ab}{k_{-1}k_{-2}} + \dfrac{k_{-4}q}{k_{+4}}}$$

which may be substituted into equation 6.12 to give

$$v^* \; = \; \frac{k_{+1}k_{+2}k_{+3}e_0 a^* b}{\left(1 + \dfrac{k_{+1}a}{k_{-1}} + \dfrac{k_{+1}k_{+2}ab}{k_{-1}k_{-2}} + \dfrac{k_{-4}q}{k_{+4}}\right)[k_{-1}(k_{-2} + k_{+3}) + k_{+2}k_{+3}b]}$$

$$(6.13)$$

This equation does not contain p because, if equilibrium is to be maintained, only three of the four reactant concentrations can be chosen at will. Any one of a, b and q can be replaced with p by means of the identity

$$K \; = \; \frac{k_{+1}k_{+2}k_{+3}k_{+4}}{k_{-1}k_{-2}k_{-3}k_{-4}} \; = \; \frac{pq}{ab}$$

If b and p are varied in a constant ratio (in order to maintain equilibrium) at fixed values of a and q, the effect on the exchange rate can be seen by realizing that the denominator of equation 6.13 is a quadratic in b, whereas the numerator is directly proportional to b. Hence the equation has the same form as that for simple substrate inhibition (Section 5.11). It follows therefore that as b and p are increased from zero to saturation, the rate of exchange increases to a maximum and then decreases to zero.

The equations for any other exchange reaction can be derived in a similar manner. In the compulsory-order ternary-complex mechanism, exchange from B* to P* or Q* is not inhibited by A, because saturating concentrations of A do not remove EA, but instead bring its concentration to a maximum. Similar results apply to the reverse reaction: exchange from Q* is inhibited by excess of P, but exchange from P* is not inhibited by excess of Q.

The random-order ternary-complex mechanism differs from the compulsory-order mechanism in that no exchange can be completely inhibited by the alternate substrate. For example, if B is present in excess, the pathway for A* to P* exchange discussed above is inhibited because E is removed from the system, but the exchange is not prevented completely because an alternative pathway is available: at high concentrations of B, A* can enter into exchange reactions by binding to EB to give EA*B. As radioactive counting can be made very sensitive, it is possible to detect very minor alternative pathways by isotope exchange. The warning must be given, however, that isotope-exchange experiments require more highly purified enzyme than conventional kinetic experiments if valid results are to be obtained. The reason for this requirement is very simple. Suppose one is studying alcohol dehydrogenase, which catalyses the reaction

$$\text{ethanol} + NAD^+ \rightleftharpoons \text{acetaldehyde} + NADH$$

Small amounts of contaminating enzymes, for example other NAD-dependent dehydrogenases, are of little importance if one is following the complete reaction, because it is unlikely that any of the contaminants is a catalyst for the complete reaction. Exchange between NAD^+ and NADH is another matter, however, and one must be certain that contaminating dehydrogenases are not present if one wants to obtain valid information about alcohol dehydrogenase.

Isotope exchange permits a useful simplification of the substituted-enzyme mechanism, in that one can study one half of the reaction only:

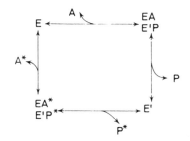

This mechanism is of the same form as the complete mechanism, with P* and A* replacing B and Q, respectively, but the kinetics are simpler because the rate constants are the same for the two halves of the reaction. This type of exchange represents a major qualitative difference between the substituted-enzyme and ternary-complex mechanisms, because in ternary-complex mechanisms no exchange can occur unless the system is complete. This method of distinguishing between the two types of mechanism was, in fact, used and discussed

(Doudoroff, Barker and Hassid, 1947; Koshland, 1955) well before
the introduction of isotope exchange as a kinetic technique.

The possibility of studying only parts of mechanisms in this way
is particularly valuable with more complex substituted-enzyme mechanisms with three or more substrates. In such cases, any simplification
of the kinetics is obviously to be welcomed, and this approach has
been used with some success, for example by Cedar and Schwartz
(1969) in the study of asparagine synthetase.

6.9 Reactions with three or more substrates

Reactions with three or more substrates can be studied by methods
that are a logical extension of those described earlier in this chapter,
and for that reason they do not require as much detailed discussion
as two-substrate reactions. Nonetheless they are not uncommon or
unimportant in biochemistry, and in this section I shall outline some
of the main points, with particular attention to characteristics that
are not well exemplified by two-substrate kinetics.

Three-substrate reactions do not necessarily have three products —
indeed, reactions with three substrates and two products are common —
but to keep the discussion within manageable limits I shall consider
only a reaction with three substrates, A, B and C, and three products,
P, Q and R. If the mechanism contains branched pathways, the complete steady-state rate equation contains terms in the squares and
possibly higher powers of the reactant concentrations, but if no such
higher-order dependence is observed, the most general equation for
the initial rate in the absence of products is

$$v = \frac{Vabc}{K^{ABC} + K^{BC}a + K^{AC}b + K^{AB}c + K_m^C ab + K_m^B ac + K_m^A bc + abc}$$

$$(6.14)$$

in which V is the limiting rate when all three substrates are extrapolated
to saturation; K_m^A, K_m^B and K_m^C are the Michaelis constants for the
three substrates, that is, the apparent values when the other two substrates are extrapolated to saturation; and K^{ABC}, K^{BC}, K^{AC} and K^{AB}
are products of Michaelis and other constants with specific meanings
that depend on the particular mechanism considered, but which are
analogous to the product $K_i^A K_m^B$ that occurs in equation 6.5.

Equation 6.14 applies in full if the reaction proceeds through a
quaternary complex EABC that exists in the steady state in equilibrium
with the free enzyme E and all possible binary and ternary complexes,
i.e. EA, EB, EC, EAB, EAC and EBC. In addition to this fully random-
order rapid-equilibrium mechanism, a range of other quaternary-complex mechanisms are possible, in which the order of binding is fully

or partly compulsory. The extreme case is the fully compulsory-order mechanism, in which there is only one binary complex, say EA, and only one ternary complex, say EAB, possible between E and EABC. Plausible intermediate cases are ones in which there are two binary complexes and one ternary complex, say EA, EB and EAB, or one binary complex and two ternary complexes, say EA, EAB and EAC.

The classification of two-substrate mechanisms into ternary-complex and substituted-enzyme mechanisms also has its parallel for three-substrate mechanisms, but again the range of possibilities is considerably greater. The extreme type of substituted-enzyme mechanism is one in which only binary complexes occur and each substrate-binding step is followed by a product-release step; in addition, a three-substrate three-product reaction may combine features of both kinds of mechanism. For example, in a common type of mechanism two substrate molecules may bind to form a ternary complex, but the first product is released before the third substrate binds.

It will be clear that the number of conceivable mechanisms is extremely large, and even if chemically implausible ones are excluded (as is not always done) there are still about 18 reasonable three-substrate three-product mechanisms (listed, for example, by Wong and Hanes, 1969), without considering such complexities as non-productive complexes and isomerizations. It is thus especially important to consider chemical plausibility in studying the kinetics of three-substrate reactions. Moreover, provided the rate appears to obey the Michaelis–Menten equation for each substrate considered separately, it is usual practice to use rate equations derived on the assumption that random-order portions of mechanisms are at equilibrium, whereas compulsory-order portions are in a steady state. This, of course, prevents the appearance of higher-order terms in the rate equation and provides much scope for the use of Cha's method (Section 4.3).

Kinetically, the various mechanisms differ in that they generate equations similar to equation 6.14 with some of the denominator terms missing, as first noted by Frieden (1959). For example, with the following mechanism:

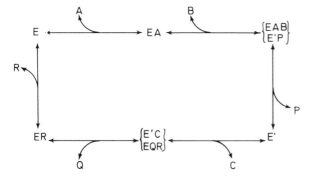

it is evident from inspection that the constant and the terms in a and b are missing from the denominator of the rate equation (because one cannot find any King–Altman patterns that contain no concentrations, or a only, or b only; cf. Section 4.4). Thus instead of equation 6.14 the rate with this mechanism is given by

$$v = \frac{Vabc}{K^{AB}c + K_m^C ab + K_m^B ac + K_m^A bc + abc} \tag{6.15}$$

For any substrate varied at constant concentrations of the other two this equation is of the form of the Michaelis–Menten equation, with

TABLE 6.4   Apparent constants for an example of a three-substrate mechanism

The table gives expressions for the apparent values of the parameters of the Michaelis–Menten equation for a three-substrate reaction that obeys equation 6.15.

Variable substrate	V^{app}	V^{app}/K_m^{app}	K_m^{app}
A	$\dfrac{Vbc}{K_m^C b + K_m^B c + bc}$	$\dfrac{Vb}{K^{AB} + K_m^A b}$	$\dfrac{(K^{AB} + K_m^A b)c}{K_m^C b + K_m^B c + bc}$
B	$\dfrac{Vac}{K_m^C a + K_m^A c + ac}$	$\dfrac{Va}{K^{AB} + K_m^B a}$	$\dfrac{(K^{AB} + K_m^B a)c}{K_m^C a + K_m^A c + ac}$
C	$\dfrac{Vab}{K^{AB} + K_m^B a + K_m^A b + ab}$	V/K_m^C	$\dfrac{K_m^C ab}{K^{AB} + K_m^B a + K_m^A b + ab}$

apparent constants as listed in *Table 6.4*. As usual, the behaviour of K_m^{app} is too complicated to be directly useful, but the other two parameters are informative. I shall here discuss only the behaviour of V^{app}/K_m^{app}, but it is also instructive to examine the expressions for V^{app} and compare them with equations 6.5 and 6.9. For variable a, V^{app}/K_m^{app} increases with b but is independent of c; for variable b, V^{app}/K_m^{app} increases with a but is independent of c; for variable c, V^{app}/K_m^{app} is a constant, independent of both a and b. This immediately distinguishes A and B from C but not from each other. A and B can be distinguished, however, by considering the effect of adding a single product.

Although the rate equation contains no term in a alone, it does contain a term in ap if P is added to the reaction mixture, as one may readily confirm by inspection. Terms in aq or ar cannot, however, be generated by addition of Q or R, and none of the three products alone can generate a term in bp, bq or br. Treating p as a constant, we can, in the terminology of Wong and Hanes (1969), say that addition of P *recalls* the missing term in a to the rate equation. On the other hand, Q and R cannot recall the term in a and none of the products can recall the term in b. The practical consequence of this is that if P is present in the reaction mixture, V^{app}/K_m^{app} for variable b becomes dependent on c but V^{app}/K_m^{app} for variable a remains independent of c regardless of which product is present.

Product inhibition in three-substrate three-product reactions obeys principles similar to those outlined in Section 6.6, with the additional feature that uncompetitive inhibition becomes a relatively common phenomenon: it occurs for at least one substrate–product pair in all compulsory-order mechanisms. For the mechanism we have been discussing, for example, Q must be uncompetitive with respect to both A and B, because, in the absence of both P and R, all King–Altman patterns giving a dependence on q also include ab. Similarly, R must be uncompetitive with respect to C.

This brief discussion of some salient points of three-substrate mechanisms, with emphasis on a single example, cannot be more than an introduction to a large subject. For more information, *see* Wong and Hanes (1969) and Dalziel (1969). Dixon and Webb (1963) discuss the application of isotope exchange to three-substrate reactions, though the comments of Dalziel (1969) about possibly misleading results should be noted.

The analysis of four-substrate reactions has been outlined by Elliott and Tipton (1974), and follows principles similar to that of three-substrate reactions.

Problems

6.1 The progressive hydrolysis of the $\alpha(1\rightarrow 4)$ glucosidic bonds of amylose is catalysed both by α-amylase and by β-amylase. In the case of α-amylase, the newly formed reducing group has the same α-configuration (before mutarotation) as the corresponding linkage in the polymer, whereas in the case of β-amylase it has the β-configuration. Suggest reasonable mechanisms for the two group-transfer reactions that would account for these observations.

6.2 Petersen and Degn (1978) have reported that when laccase from *Rhus vernicifera* catalyses the oxidation of hydroquinone by molecular oxygen, the rate increases indefinitely as the concentrations of both substrates are increased in concert, with no evidence of saturation. They account for these observations in terms of a substituted-enzyme mechanism in which the initial oxidation of the enzyme by oxygen occurs in a single step, followed by a second step in which the original form of the enzyme is regenerated as a result of reduction of the oxidized enzyme by hydroquinone. Explain why this mechanism accounts for the inability of the substrates to saturate the enzyme.

6.3   Sucrose glucosyltransferase catalyses the reaction

glucose 1-phosphate + fructose ⇌ sucrose + inorganic phosphate

In the absence of both sucrose and fructose, the enzyme also catalyses rapid ^{32}P exchange between glucose 1-phosphate and labelled inorganic phosphate. The exchange is strongly and competitively inhibited by glucose. The enzyme is a rather poor catalyst for the hydrolysis of glucose 1-phosphate. How may these results be explained?

6.4   Derive an equation for the initial rate in the absence of added products of a reaction obeying a compulsory-order ternary-complex mechanism, with A binding first and B binding second, assuming that both substrate-binding steps are at equilibrium. How does the equation differ in form from the ordinary steady-state equation for this mechanism? What would be the appearance of primary plots of b/v against b?

6.5   The rate of an enzyme-catalysed reaction with two substrates is measured with the two concentrations a and b varied at a constant value of a/b. What would be the expected shape of a plot of a/v against a if the reaction followed (a) a ternary-complex mechanism? or (b) a substituted-enzyme mechanism?

6.6   What set of product-inhibition patterns would be expected for an enzyme that obeyed a rapid-equilibrium random-order ternary-complex mechanism?

6.7   Consider a reaction with substrates A and B that follows a substituted-enzyme mechanism. Without deriving a complete rate equation, determine the type of inhibition expected for an inhibitor that binds in a dead-end reaction to the form of the free enzyme that binds B, but has no effect on the other form of the free enzyme.

6.8   In the symbolism of Dalziel (1957), which is often encountered in the literature, equation 6.5 would take the form

$$\frac{e_0}{v} = \phi_0 + \frac{\phi_1}{S_1} + \frac{\phi_2}{S_2} + \frac{\phi_{12}}{S_1 S_2}$$

in which e_0 and v have the same meanings as in this book, S_1 and S_2 represent a and b respectively, and ϕ_0, ϕ_1, ϕ_2 and ϕ_{12} are constants, sometimes known as Dalziel coefficients. What are the values of these constants in terms of the symbols used in equation 6.5? At what point (expressed in terms of Dalziel coefficients) do the straight lines obtained by plotting S_1/v against S_1 at different values of S_2 intersect?

6.9 Consider a three-substrate three-product reaction as follows:

$$A + B + C \rightleftharpoons P + Q + R$$

that proceeds by a quaternary-complex mechanism in which the substrates bind and the products are released in the order shown in the equation. The initial rate in the absence of products is given by equation 6.14 with one term missing.

(a) Which term is missing?

(b) Which product, if any, can 'recall' this term to the rate equation?

(c) Which product acts as an uncompetitive inhibitor (in the absence of the other two products) regardless of which substrate concentration is varied?

(d) Which product behaves as a competitive inhibitor when A is the variable substrate?

[*Note.* This problem can be solved by inspection (cf. Section 4.4); it is not necessary to derive a complete rate equation.]

Chapter 7

Effects of pH and temperature on enzymes

7.1 pH and enzyme kinetics

Of the many problems that beset the earliest investigators of enzyme kinetics, none was more important than the lack of understanding of hydrogen-ion concentration, $[H^+]$. In aqueous chemistry, $[H^+]$ varies from about 1 M to about 10^{-14} M, an enormous range that is commonly decreased to more manageable proportions by the use of a logarithmic scale, $pH = -\log [H^+]$. All enzymes are profoundly influenced by pH, and no substantial progress could be made in the understanding of enzymes until Michaelis and his collaborators made pH control a routine characteristic of all serious enzyme studies. The concept of buffers for controlling the hydrogen-ion concentration, and the pH scale for expressing it, were first published by Sørensen (1909), in a classic paper on the importance of hydrogen-ion concentration in enzymic studies. It is clear (*see* Michaelis, 1958), however, that Michaelis was already working on similar lines, and it was not long afterwards that the first of his many papers on pH effects on enzymes appeared (Michaelis and Davidsohn, 1911). Although there are still some doubts about the proper interpretation of pH effects in enzyme kinetics, the practical importance of pH continues undiminished: it is hopeless to attempt any kinetic studies without adequate control of pH.

It is perhaps surprising that it was left to enzymologists to draw attention to the importance of hydrogen-ion concentration and to introduce the use of buffers. It is worth while therefore to reflect on the special properties of enzymes that made pH control imperative before any need for it had been felt in the already highly developed science of chemical kinetics. With a few exceptions, such as pepsin and alkaline phosphatase, the enzymes that have been most studied are active only in aqueous solution at pH values in the range 5–9. Indeed, only pepsin has a physiologically important activity outside this middle range of pH. Now, in the pH range 5–9, the hydrogen-ion and hydroxide-ion concentrations are both in the range 10^{-5}–10^{-9} M, i.e. very low, and are very sensitive to impurities. Whole-cell extracts, and crude enzyme preparations in general, are well buffered by enzyme and other polyelectrolyte impurities, but this natural buffering is lost when an enzyme is purified, and must be replaced with artificial buffers. Until this effect was realized, little progress in enzyme kinetics was possible. This situation can be contrasted with that in general chemistry: only a minority of reactions are studied in aqueous solution and, of these, the majority are studied either at very high or at very low pH, at which the concentration of either hydrogen or hydroxide ion is high enough to be reasonably stable. Consequently, the early development of chemical kinetics was little hampered by the lack of understanding of pH.

The simplest type of pH effect on an enzyme, when only a single acidic or basic group is involved, is no different from the general case of hyperbolic inhibition and activation that was considered in Section 5.8. Conceptually, the protonation of a basic group on an enzyme is simply a special case of the binding of a modifier at a specific site and there is therefore no need to repeat the algebra for this simplest case. However, there are several differences between protons and other modifiers that make it worth while to examine protons separately. First, virtually all enzymes are affected by protons, so that the proton is far more important than any other modifier. It is far smaller than any other chemical species and has no steric effect; this means that certain phenomena, such as pure non-competitive inhibition, are common with the proton as inhibitor but very rare otherwise. The proton concentration can be measured and controlled over a range that is enormously greater than that available for any other modifier and therefore one can expect to be able to observe any effects that might exist. Finally, protons normally bind to many sites on an enzyme, so that it is often insufficient to consider binding at one site only.

7.2 Acid–base properties of proteins

Of the various definitions of acids and bases that the student of chemistry encounters, by far the most important in enzymology is that of Brønsted (1923): 'An acid is a species having a tendency to lose a proton, and a base is a species having a tendency to add on a proton'. Apart from its emphasis on the proton, this definition is noteworthy in that it refers to *species*, which include ions as well as molecules. Unfortunately, biochemists have conventionally classified the ionizable groups found in proteins according to the properties of the amino acids in the pure uncharged state. Accordingly, aspartate and glutamate, which are largely responsible for the *basic* properties of proteins under physiological conditions, are commonly referred to as 'acidic'. Of the so-called 'basic' amino acids, histidine can act either as a base or as an acid under physiological conditions, lysine exists primarily as an acid, and arginine is largely irrelevant to the acid–base properties of proteins, because it does not deprotonate below pH 12. An attempt at a more rational classification is shown in *Table 7.1*. This has almost nothing in common with that found in most general biochemistry textbooks, but is instead based on the Brønsted definition.

TABLE 7.1   Ionizable groups in proteins

Name	Type of group	pK_a†	Brønsted character at pH 7.0
C-terminal	carboxylate	3.4	basic
Aspartate	carboxylate	4.1	basic
Glutamate	carboxylate	4.5	basic
Histidine	imidazole	6.3	mainly basic
N-terminal	amine	7.5	mainly acidic
Cysteine	thiol	8.3	acidic
Tyrosine	phenol	9.6	acidic
Lysine	amine	10.4	acidic

†pK_a values are average values at 25 °C for groups in 'typical' environments in proteins, and are based on values given by Steinhardt and Reynolds (1969). Individual groups in special environments may be 'perturbed', i.e. they may have pK_a values substantially different from those given here.

Some of the groups included in *Table 7.1*, such as the C-terminal carboxylate and the ε-amino group of lysine, have pK_a values so far from 7 that it might seem unlikely that they would contribute to the catalytic properties of enzymes. The values given in the table are average values for groups in 'typical' environments, however, and may differ substantially from the pK_a values exhibited by individual groups in special environments, such as the vicinity of the active site. Such pK_a values are said to be *perturbed*. A clear-cut example is provided by pepsin, which has an isoelectric point of 1.4. As there are four groups that are presumably cationic at low pH, there must be at least four groups with pK_a values well below the range expected for carboxylic acids. Although the enzyme contains a phosphorylated serine residue, this can only partly account for the low isoelectric point, and there must inevitably be at least three perturbed carboxylic acid groups. A possible explanation is that if two acidic groups are held in close proximity one would expect the singly protonated state to be stabilized with respect to the doubly protonated and doubly deprotonated states (*see* Knowles *et al.*, 1970).

7.3 Ionization of a dibasic acid

The pH behaviour of many enzymes can be interpreted as a first approximation in terms of a simple model due to Michaelis (1926), in which only two ionizable groups are considered. The enzyme may be represented as a dibasic acid, HEH, with two non-identical acidic groups:

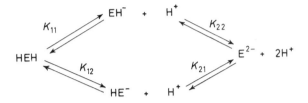

With the dissociation constants defined as shown in this scheme, the concentrations of all forms of the enzyme can be represented at equilibrium in terms of the hydrogen-ion concentration, $[H^+]$, or, more conveniently, h:

$$[EH^-] = [HEH]K_{11}/h$$

$$[HE^-] = [HEH]K_{12}/h$$

$$[E^{2-}] = [HEH]K_{11}K_{22}/h^2 = [HEH]K_{12}K_{21}/h^2 \qquad (7.1)$$

Two points should be noted about these relationships: first, although K_{11} and K_{21} both define the dissociation of a proton from the same group, HE^- is more negative than HEH by one unit of charge and so one would expect it to be less acidic, i.e. $K_{11} > K_{21}$, not $K_{11} = K_{21}$; similarly, $K_{12} > K_{22}$. Second, the concentration of E^{2-} must be the same whether it is derived from HEH via EH^- or via HE^-; the two expressions for $[E^{2-}]$ in equation 7.1 must therefore be equivalent, i.e. $K_{11}K_{22} = K_{12}K_{21}$.

If the total enzyme concentration is $e_0 = [HEH] + [EH^-] + [HE^-] + [E^{2-}]$, then

$$[HEH] = \frac{e_0}{1 + \dfrac{K_{11} + K_{12}}{h} + \dfrac{K_{11}K_{22}}{h^2}} \qquad (7.2)$$

$$[EH^-] = \frac{e_0 K_{11}/h}{1 + \dfrac{K_{11} + K_{12}}{h} + \dfrac{K_{11}K_{22}}{h^2}} \qquad (7.3)$$

$$[HE^-] = \frac{e_0 K_{12}/h}{1 + \dfrac{K_{11} + K_{12}}{h} + \dfrac{K_{11}K_{22}}{h^2}} \qquad (7.4)$$

$$[E^{2-}] = \frac{e_0 K_{11} K_{22}/h^2}{1 + \dfrac{K_{11} + K_{12}}{h} + \dfrac{K_{11} K_{22}}{h^2}} \qquad (7.5)$$

These expressions show how the concentrations of the four species vary with h, and, by extension, with pH, and a typical set of curves is shown in *Figure 7.1*, with arbitrary values assumed for the dissociation constants. In a real experiment, one can never define the curves

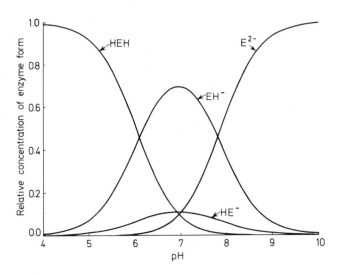

Figure 7.1   Relative concentrations of enzyme forms as a function of pH, for an enzyme HEH with two ionizable groups: $pK_{11} = 6.1$; $pK_{12} = 6.9$; $pK_{21} = 7.0$; $pK_{22} = 7.8$

as precisely as this, because one cannot evaluate the four dissociation constants. The reason for this can be seen by considering the fact that $[EH^-]/[HE^-] = K_{11}/K_{12}$, that is, a constant, independent of h. Thus no amount of variation of h will produce any change in $[EH^-]$ that is not accompanied by an exactly proportional change in $[HE^-]$. Consequently, it is impossible to determine how much of any given property is contributed by EH^- and how much by HE^- and for practical purposes we must therefore treat EH^- and HE^- as a single species, with concentration given by

$$[EH^-] + [HE^-] = e_0 \bigg/ \left(\frac{h}{K_1} + 1 + \frac{K_2}{h}\right) \qquad (7.6)$$

where $K_1 = K_{11} + K_{12} = ([EH^-] + [HE^-])h/[HEH]$ and $K_2 = K_{11} K_{22}/(K_{11} + K_{12}) = [E^{2-}]h/([EH^-] + [HE^-])$. K_1 and K_2 are called *molecular dissociation constants*, to distinguish them from K_{11}, K_{12}, K_{21} and K_{22}, which are *group dissociation constants*. They have the practical advantage that they can be measured, whereas the conceptually preferable group dissociation constants cannot, because it is not possible to evaluate K_{12}/K_{11}.

The expressions for [HEH] and [E^{2-}] can also be written in terms of molecular dissociation constants:

$$[\text{HEH}] = e_0 \left/ \left(\frac{h^2}{K_1 K_2} + \frac{h}{K_2} + 1 \right) \right.$$

$$[\text{E}^{2-}] = e_0 \left/ \left(1 + \frac{K_1}{h} + \frac{K_1 K_2}{h^2} \right) \right.$$

We shall now examine equation 7.6 in more detail, because many enzymes display the bell-shaped pH profile characteristic of this equation. The curves for EH$^-$ and HE$^-$ in *Figure 7.1* are of this form, and a representative set of bell-shaped curves for different values of (pK_2 − pK_1) is given in *Figure 7.2*. Notice that the curves

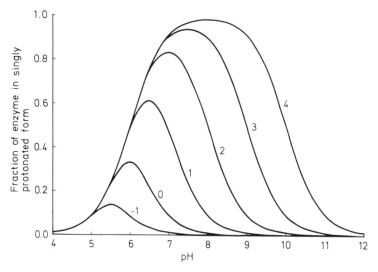

Figure 7.2 Bell-shaped curves calculated from equation 7.6, with p$K_1 = 6.0$ and p$K_2 = 5.0$–10.0. Each curve is labelled with the value of pK_2 − pK_1, as this quantity determines its shape

are not all the same shape: the maximum becomes noticeably flat as (pK_2 − pK_1) increases. The curve does not approach a maximum of 1.0 unless (pK_2 − pK_1) is greater than about 3. Consequently the values of the pH at which [EH$^-$] + [HE$^-$] is half-maximal are not equal to pK_1 and pK_2. However, the mean of these two pH values *is* equal to ½(pK_1 + pK_2), and is also the pH at which the maximum occurs. The relationship between the width at half height of the curve and (pK_2 − pK_1) is shown in *Table 7.2*. This table allows measurements of the pH values where the ordinate is half-maximal

TABLE 7.2 Relationship between the width at half height and the pK difference for bell-shaped pH profiles

The most convenient method for calculating the pK difference from the width at half height is the following, suggested by Dixon (1979): define the width at half height as $2 \log q$; then $pK_2 - pK_1 = 2 \log (q - 4 + 1/q)$. Alternatively, if pH_1 and pH_2 are the pH values at which the measured parameter is half-maximal, and are given by $pH_{max} \mp \log q$, where pH_{max} is the pH at the maximum, then pK_1 and pK_2 are given by $pH_{max} \mp \log (q - 4 + 1/q)$.

Width at half height	$pK_2 - pK_1$	Width at half height	$pK_2 - pK_1$	Width at half height	$pK_2 - pK_1$
1.14[†]	$-\infty$	2.1	1.73	3.1	3.00
1.2	-1.27	2.2	1.88	3.2	3.11
1.3	-0.32	2.3	2.02	3.3	3.22
1.4	0.17	2.4	2.15	3.4	3.33
1.5	0.51	2.5	2.28	3.5	3.44
1.6	0.78	2.6	2.41	3.6	3.54
1.7	1.02	2.7	2.53	3.7	3.65
1.8	1.22	2.8	2.65	3.8	3.76
1.9	1.39	2.9	2.77	3.9	3.86
2.0	1.57	3.0	2.88	4.0	3.96

†The width at half height of a curve defined by equation 7.6 cannot be less than 1.14.

to be converted into molecular pK values. Nonetheless, it must be remembered that even if pK_1 and pK_2 are correctly estimated, the values of the group dissociation constants remain unknown, unless plausibility arguments are invoked, with unprovable assumptions (Dixon, 1976).

7.4 Effect of pH on enzyme kinetic constants

By a simple extension of the theory for the ionization of a dibasic acid, one can account for the bell-shaped activity curves that are often observed for the enzyme kinetic constants V and V/K_m. (The treatment of K_m is more complex, as we shall see.) The basic mechanism is as follows:

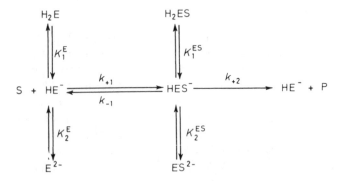

The free enzyme is again treated as a dibasic acid, $H_2 E$, with two molecular dissociation constants, K_1^E and K_2^E, as in the previous section, and the enzyme–substrate complex $H_2 ES$ is similar, but with dissociation constants K_1^{ES} and K_2^{ES}. Only the singly ionized complex, HES^-, is able to react to give products. Before proceeding further, I must emphasize that this scheme includes several implied assumptions that may be oversimplifications. First, the omission of substrate-binding steps for $H_2 E$ and E^{2-} implies that the protonation steps are dead-end reactions, so that they can be treated as equilibria (*see* Section 4.4). However, this is nothing more than begging the question, because in most cases it is most unreasonable to postulate that S cannot bind directly to $H_2 E$ and E^{2-}. If these steps are included, the protonation steps cease to be dead-end reactions, and can then be treated as equilibria only if they are assumed to be very rapid compared with other steps. This may seem to be a reasonable assumption, in view of the simple nature of the reaction, but it may not always be true, particularly if protonation is accompanied by a compulsory conformation change.

The scheme also implies that the catalytic reaction involves only two steps, as in the simplest Michaelis–Menten mechanism. If several steps are postulated, with each intermediate capable of protonation and deprotonation, the form of the final equation is not affected, but it is no longer possible to interpret experimental results in a straightforward and simple way. (Compare the effect of introducing an extra step into the simple Michaelis–Menten mechanism, Section 2.6.)

Finally, the assumption that only HES^- can break down to give products may not always be true, but it is likely to be reasonable for many enzymes because most enzyme activities do approach zero at high and low pH values.

Recognizing that the scheme given at the beginning of this section may be an optimistic representation of the actual situation, let us consider the rate equation that it predicts. If there were no ionizations, and HE^- and HES^- were the only forms of the enzyme, then the scheme would simplify to the ordinary Michaelis–Menten mechanism, with a rate given by

$$ v = \frac{k_{+2}e_0 s}{(k_{-1} + k_{+2})/k_{+1} + s} = \frac{\tilde{V}s}{\tilde{K}_m + s} $$

where $\tilde{V} = k_{+2}e_0$ and $\tilde{K}_m = (k_{-1} + k_{+2})/k_{+1}$ are the *pH-corrected* constants, convenient fictions analogous to the *expected values* discussed in the context of non-productive binding (Section 5.10). In reality, however, the free enzyme does not exist solely as HE^-, nor the enzyme–substrate complex solely as HES^-. The full rate equation is of exactly the same form, i.e.

$$v = \frac{Vs}{K_m + s}$$

but the parameters V and K_m are not equal to $\tilde{V}$ and $\tilde{K}_m$; instead they are functions of h, and the expressions for V and V/K_m are of the same form as equation 7.6:

$$V = \tilde{V}/[(h/K_1^{ES}) + 1 + (K_2^{ES}/h)] \tag{7.7}$$

$$V/K_m = (\tilde{V}/\tilde{K}_m)/[(h/K_1^E) + 1 + (K_2^E/h)] \tag{7.8}$$

V reflects the ionization of the enzyme–substrate complex; V/K_m reflects the ionization of the free enzyme (or the free substrate; *see* Section 7.5). In either case the pH dependence follows a symmetrical bell-shaped curve of the type discussed in the previous section.

The variation of K_m with pH is more complicated:

$$K_m = \tilde{K}_m [(h/K_1^E) + 1 + (K_2^E/h)]/[(h/K_1^{ES}) + 1 + (K_2^{ES}/h)] \tag{7.9}$$

as it depends on all four pK values. Nonetheless, it is possible in principle to obtain all four pK values by plotting log K_m against pH and applying a theory developed by Dixon (1953a). It is obvious from inspection that at high h (low pH) equation 7.7 simplifies to $V = \tilde{V}h/K_1^{ES}$ and that at low h (high pH) it simplifies to $V = \tilde{V}K_2^{ES}/h$. If K_1^{ES} and K_2^{ES} are well separated, there is also an intermediate region in which $V \simeq \tilde{V}$. It follows that a plot of log V against pH should approximate to the form shown in *Figure 7.3a*, with three straight-line sections intersecting at pH = pK_1^{ES} and pH = pK_2^{ES}. The behaviour of V/K_m, shown in *Figure 7.3b*, is similar, except that the intersections occur at pH = pK_1^E and pH = pK_2^E. In spite of the complexity of equation 7.9, the form of a plot of log K_m against pH follows simply from the fact that $K_m = V/(V/K_m)$ and so log K_m = log V − log (V/K_m). Accordingly, the plot shown in *Figure 7.3c* is the result of subtracting the ordinate values of the line in *Figure 7.3b* from those in *Figure 7.3a*. This plot approximates to a series of straight lines of slope +1, 0 or −1 (slopes of +2 and −2 are also possible, though they do not occur in *Figure 7.3c*): as one reads across the graph from left to right, each increase in slope corresponds to a pK on the free enzyme, pK_1^E or pK_2^E, and each decrease in slope corresponds to a pK on the enzyme–substrate complex, pK_1^{ES} or pK_2^{ES}.

The plots shown in *Figure 7.3* are idealized, of course, and it would be very unusual to have accurate data available for a wide enough

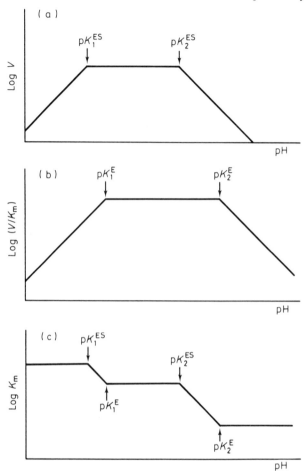

Figure 7.3 Interpretation of pH profiles according to the theory of Dixon (1953a). In reality, plots of log (parameter) against pH should always give smooth curves, but these approximate to the sets of straight-line segments illustrated. (a) Changes in slope on the plot of log V against pH reflect ionizations of the enzyme–substrate complex; (b) changes in slope on the plot of log (V/K_m) against pH reflect ionizations of the free enzyme; (c) in principle, all ionizations affect K_m, in a way that can be rationalized by regarding log K_m as log V − log (V/K_m)

range of pH to provide all four pK values. However, the interpretation of changes in slope is the same even if only part of the plot is available.

In concluding this section I should emphasize that pH-dependence curves should refer to the parameters of the Michaelis–Menten equation, V and V/K_m. In other words a *series* of initial rates should be measured at *each* pH value, so that V and V/K_m can be determined at each pH value. The pH dependence of v is of little use by itself because competing effects on V and V/K_m can make any pK_a values apparently measured highly misleading. In this respect measurements of the effects of pH should follow the same principles as measurements of the effects of changes in other environmental influences on enzymes, such as temperature, ionic strength, concentrations of inhibitors and activators, etc.

7.5 Ionization of the substrate

Many substrates ionize in the pH range used in kinetic experiments. If substrate ionization is possible one should therefore consider whether observed pK values refer to the enzyme or the substrate. The theory of this case is similar to that for enzyme ionization and the results given above require only slight modification. The pH dependence of V, and decreases in slope in plots of log K_m against

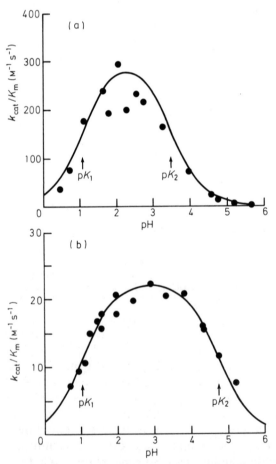

Figure 7.4   pH dependence of k_{cat}/K_m for the pepsin-catalysed hydrolysis of (a) acetyl-L-phenylalanyl-L-phenylalanylglycine, and (b) acetyl-L-phenylalanyl-L-phenylalaninamide (Cornish-Bowden and Knowles, 1969)

pH, still refer to the enzyme–substrate complex; but the pH dependence of V/K_m, and increases in slope on the plot of log K_m against pH, may refer *either* to the free enzyme *or* to the free substrate. In some cases one may be able to decide which interpretation is correct by studying another substrate that does not ionize. For example (*Figure 7.4*), the pH dependence of k_{cat}/K_m for the pepsin-catalysed hydrolysis of acetyl-L-phenylalanyl-L-phenylalanylglycine shows pK values of 1.1 and 3.5, of which the latter may well be due to ionization of the substrate. That this interpretation is correct is confirmed by consideration of a substrate that does not ionize, acetyl-L-phenylalanyl-L-phenylalaninamide: this showed essentially the same pK_1, 1.05, but a higher value of pK_2, 4.75 (*Figure 7.4b*), which presumably refers to an ionization of the enzyme.

7.6 More complex pH effects

One of the principal reasons why pH-dependence studies are made is to measure pK values and deduce from them the chemical nature of the groups on the enzyme that participate in catalysis. Although this is widely done, it demands more caution than is sometimes apparent, because simple treatments of pH effects make a variety of assumptions that may not always be valid. Knowles (1976) has critically discussed the assumptions that are commonly made in interpreting pH profiles and has shown how one can be led to false conclusions; his paper should be read by anyone with a serious interest in pH effects on enzymes.

It is not only the quantitative assumptions that are suspect; the qualitative interpretation of a pH profile can also be misleading. For example, a bell-shaped pH profile may well indicate a requirement for two groups in the system to exist in particular ionic states, as discussed in Sections 7.4 and 7.5, but this is not the only possibility: in some circumstances a single group that is required in different states for two steps of the reaction may give similar behaviour (Dixon, 1973; Cornish-Bowden, 1976). This is an example of the more general phenomenon of a change of rate-determining step with pH (Jencks, 1969). For a fuller discussion of this and other more complex pH effects than is possible in this chapter, *see* Tipton and Dixon (1979).

7.7 Temperature dependence of enzyme-catalysed reactions

In principle, the theoretical treatment discussed in Sections 1.6 and 1.7 for the temperature dependence of simple chemical reactions applies equally well to enzyme-catalysed reactions, but in practice several complications arise that must be properly understood if any useful information is to be obtained from temperature-dependence studies. First, almost all enzymes become denatured if they are heated much above physiological temperatures, and the conformation of the enzyme is altered, often irreversibly, with loss of catalytic activity. Denaturation is chemically a very complex and only partly understood process, and only a simplified account will be given here: I shall consider reversible denaturation only and I shall assume that an equilibrium exists at all times between the active and denatured enzyme and that only a single denatured species need be considered.

Denaturation does not involve rupture of covalent bonds, but only of hydrogen bonds and other weak interactions that are involved in maintaining the active conformation of the enzyme. Although each individual hydrogen bond is far weaker than a covalent bond (about 20 kJ mol^{-1} for a hydrogen bond compared with about 400 kJ mol^{-1} for a covalent bond), denaturation generally involves the rupture of a large number of them. The standard enthalpy of reaction, ΔH^0, for denaturation is therefore often very high, typically 200–500 kJ mol^{-1}. However, the rupture of a large number of weak bonds greatly increases the number of conformational states available to an enzyme molecule and so denaturation is also characterized by a very large standard entropy of reaction, ΔS^0.

The effect of denaturation on observed enzymic rate constants can be seen by considering the simple example of an active enzyme E in equilibrium with an inactive form E′:

$$
\begin{array}{c}
\text{E}' \\
K \updownarrow \\
\text{E} + \text{S} \xrightarrow{\;k\;} \text{E} + \text{P}
\end{array}
$$

For simplicity, the catalytic reaction is represented as a simple second-order process with rate constant k, as is usually observed at very low substrate concentrations. The equilibrium constant for denaturation, K, varies with temperature according to the van't Hoff equation (Section 1.6):

$$-RT \ln K \;=\; \Delta G^0 \;=\; \Delta H^0 - T \, \Delta S^0$$

where R is the gas constant, T is the absolute temperature and ΔG^0, ΔH^0 and ΔS^0 are the standard Gibbs free energy, enthalpy and entropy of reaction, respectively. This relationship can be rearranged to provide an expression for K:

$$K \;=\; \exp\left[(\Delta S^0 / R) - (\Delta H^0 / RT)\right]$$

The rate constant k for the catalytic reaction is governed by the integrated Arrhenius equation:

$$k \;=\; A \exp\left(-E_\mathrm{a}/RT\right)$$

where A is a constant and E_a is the Arrhenius activation energy. The rate of the catalytic reaction is given by $v = k[E][S]$, but for practical application the active-enzyme concentration $[E]$ has to be expressed in terms of the total-enzyme concentration, $e_0 = [E] + [E']$, and so

$$v = ke_0 s/(1 + K)$$

The observed rate constant, k^{obs}, may be defined as $k/(1 + K)$, and varies with temperature according to the equation

$$k^{obs} = \frac{A \exp(-E_a/RT)}{1 + \exp[(\Delta S^0/R) - (\Delta H^0/RT)]}$$

At low temperatures, when $\Delta S^0/R$ is small compared with $\Delta H^0/RT$, the exponential term in the denominator is insignificant, and so k^{obs} varies with temperature in the ordinary way according to the Arrhenius equation. At temperatures above $\Delta H^0/\Delta S^0$, however, the denominator increases steeply with temperature and the rate of reaction decreases rapidly to zero.

   Although this model is oversimplified, it does show why the Arrhenius equation appears to fail for enzyme-catalysed reactions at high temperatures. In the older literature, it was common for *optimum temperatures* for enzymes to be reported, but the temperature at which k^{obs} is maximum is of no particular significance, as the temperature dependence of enzyme-catalysed reactions is often found in practice to vary with the experimental procedure. In particular, the longer a reaction mixture is incubated before analysis, the lower the 'optimum temperature' is likely to be. The explanation of this effect is that denaturation often occurs fairly slowly, so that the reaction cannot properly be treated as an equilibrium. The extent of denaturation therefore increases with the time of incubation. This ought not to be a problem with modern experimental techniques, because in continuously assayed reaction mixtures, time-dependent processes are usually obvious (*see Figure 7.5*).

   Because of denaturation, straightforward results can usually be obtained from studies of the temperature dependence of enzymes only within a fairly narrow range of temperature, say between 0 and 50 °C, but even within this range there are important hazards to be avoided. First, the temperature dependence of the initial rate commonly gives curved Arrhenius plots from which little useful information can be obtained. Such plots often show artefacts (*see*, for example, Silvius, Read and McElhaney, 1978), and a *minimum* requirement for a satisfactory temperature study is to measure a series of rates at each temperature, so that Arrhenius plots can be drawn for the separate parameters, V, K_m and V/K_m. These plots are also often curved, and there are so many possible explanations of this — for example, a change in conformation of the enzyme, a change in rate-limiting step, the existence of the enzyme as a mixture of isoenzymes, an effect of temperature on a substrate, etc. — that it is dangerous to conclude much from the shape of an Arrhenius plot unless it can be correlated with other temperature effects that can be observed independently. Massey, Curti and Ganther (1966), for example, found sharp changes in slope at about 14 °C in Arrhenius

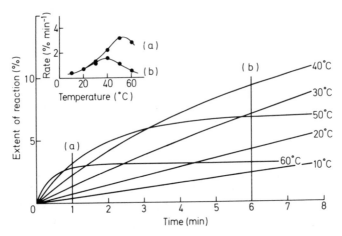

Figure 7.5 Effect of heat inactivation on the temperature dependence of the rates of enzyme-catalysed reactions. The inset shows the effect of defining the 'initial rate' as the mean rate during the first (a) 1 min, or (b) 6 min

plots for D-amino acid oxidase; at the same temperature, other techniques, such as sedimentation velocity and ultraviolet spectroscopy, indicated a change in conformational state of the enzyme. In such a case it is clearly reasonable to interpret the kinetic behaviour as a consequence of the same change in conformation.

In general one can attach little significance to studies of the temperature dependence of V, K_m or V/K_m unless the mechanistic meanings of these parameters are known. If K_m is a function of several rate constants, its temperature dependence is likely to be a complex combination of competing effects, and of little significance or interest; but if K_m is known with reasonable certainty to be a true dissociation constant, its temperature dependence can provide useful thermodynamic information about the enzyme.

Most of the 'activation energies' for enzyme-catalysed reactions that have appeared in the literature have little value, but it would be wrong to suggest that no useful information can be obtained from studies of temperature dependence; if proper care is taken very valuable information about enzyme reaction mechanisms can be obtained.

A classic study was carried out by Bender, Kézdy and Gunter (1964) on α-chymotrypsin. This work differed in almost every way from typical temperature-dependence studies, however. It included convincing evidence of the particular steps in the mechanism that were being investigated, it compared results for numerous different substrates, it referred to an enzyme about which much was known already, and it was interpreted with a proper understanding of chemistry.

Problems

7.1 For an enzyme in which K_m depends on a single ionizing group, with pK_a values pK^E in the free enzyme and pK^{ES} in the enzyme–substrate complex, equation 7.9 simplifies to $K_m = \tilde{K}_m (K^E + h)/(K^{ES} + h)$. (a) At what pH does a plot of K_m against pH show a point of inflexion? (b) At what pH does a plot of $1/K_m$ against pH show a point of inflexion? [If you find your result to be unbelievable, calculate K_m and $1/K_m$ at several pH values in the range 3–10, assuming $pK^E = 6.0$, $pK^{ES} = 7.0$, and plot both against pH. Fersht (1977) discusses the principles underlying this problem.]

7.2 Interpretation of a plot of $\log K_m$ against pH is most easily done in the light of the relationship $\log K_m = \log V - \log (V/K_m)$, in which K_m, V and V/K_m not only are dimensioned quantities, but have three different dimensions. Is this relationship a violation of the rules discussed in Section 1.3, and, if so, to what extent is the analysis implied by *Figure 7.3* invalid?

7.3 A bell-shaped pH profile has half-maximal ordinate values at pH values 5.7 and 7.5. Estimate the molecular pK_a values. Assuming that there is independent reason to believe that one of the group pK_a values is 6.1, estimate the other three.

7.4 Write down a more realistic scheme for pH dependence than that given at the beginning of Section 7.4 as follows: (a) allow both substrate and product to bind to all three forms of the free enzyme, and assume that the rate constants for these binding reactions are independent of the state of protonation; (b) assume that the catalytic process is a three-step reaction in which all steps are reversible and the second step, the interconversion of HES and HEP, occurs for the singly protonated complexes only. Assuming that all protonation reactions are at equilibrium in the steady state, use Cha's method (Section 4.3) to derive an expression for K_m as a function of the hydrogen-ion concentration. { The solution has a complicated appearance, which can be simplified by defining $f(h) = 1/[(h/K_1) + 1 + (K_2/h)]$.} Under what circumstances is K_m independent of pH? If it is independent of pH, what value must it have?

7.5   The following measurements of V for an enzyme-catalysed reaction were made over a temperature range in which no thermal inactivation could be detected. Are they consistent with interpretation of V as $k_{+2}e_0$, where e_0 is constant and k_{+2} is the rate constant for a single step in the mechanism?

Temperature (°C)	V (mM min^{-1})	Temperature (°C)	V (mM min^{-1})
5	0.32	30	11.9
10	0.75	35	19.7
15	1.67	40	30.9
20	3.46	45	46.5
25	6.68	50	68.3

Chapter 8

Control of enzyme activity

8.1 Necessity for metabolic control

It is obvious that all living organisms require a high degree of control over metabolic processes so as to permit ordered change without precipitating catastrophic progress towards thermodynamic equilibrium. It is less obvious that enzymes which behave in the way described in other chapters are unlikely to be able to provide the necessary degree of control. It is appropriate to begin therefore by examining an important step in metabolism, the interconversion of fructose 6-phosphate and fructose 1,6-bisphosphate, with a view to defining the qualities that are needed in controlled enzymes. The conversion of fructose 6-phosphate to fructose 1,6-bisphosphate requires ATP:

fructose 6-phosphate + ATP → fructose 1,6-bisphosphate + ADP

It is catalysed by phosphofructokinase and is the first step in glycolysis that is unique to glycolysis, that is, the first step that does not form part of other metabolic processes as well. It is thus an appropriate step for the control of the whole process, and there is little doubt that it is indeed the major control point. Under metabolic conditions, the reaction is essentially irreversible and, in gluconeogenesis, it is by-passed by a hydrolytic reaction, catalysed by fructose bisphosphatase:

fructose 1,6-bisphosphate + water → fructose 6-phosphate + phosphate

This reaction is also essentially irreversible. The parallel existence of two irreversible reactions is of the greatest importance in metabolic control: it means that the direction of flux can be determined by differential control of the activities of the two enzymes. A single reversible reaction could not be controlled in this way, because a catalyst cannot affect the direction of flux through a reaction, which is determined solely by thermodynamic considerations. The catalyst affects only the rate at which equilibrium can be attained.

   If both reactions were to proceed in an uncontrolled fashion at similar rates, there would be no net interconversion of fructose

6-phosphate and fructose 1,6-bisphosphate, but continuous hydroly-sis of ATP, resulting eventually in death. This situation is known as a *futile cycle*, and to prevent it it is necessary either to segregate the two processes into different cells (or different compartments of the same cell), or to control both enzymes so that each is active only when the other is inhibited. Although control is achieved by compartmen-talization to some extent, this is not possible in all circumstances, especially in tissues such as liver that can carry out both glycolysis and gluconeogenesis.

We must now consider whether an enzyme that obeys the ordinary laws of enzyme kinetics can be controlled precisely enough to prevent futile cycling. For an enzyme that obeys the Michaelis–Menten equa-tion, $v = Vs/(K_m + s)$, a simple calculation shows that the rate is $0.1V$ when $s = K_m/9$, and that it is $0.9V$ when $s = 9K_m$. In other words an enormous increase in substrate concentration, 81-fold, is required to bring about a comparatively modest increase in rate from 10% to 90% of the maximum. Similar results are obtained by considering the change in concentration of a competitive inhibitor needed to decrease the rate from 90% to 10% of the uninhibited value. Even if one considers the effect of two or more effectors acting in concert, the qualitative conclusion is the same: an inordinately large change in the environment is necessary to bring about even a modest change in rate. The requirements for control of metabolism are exactly the opposite: on the one hand, the concentrations of major metabolites must be maintained within small tolerances, and on the other hand, reaction rates must be capable of changing greatly – probably more than the $0.1V$ to $0.9V$ range we have considered – in response to fluctuations within these small tolerances.

Clearly, the ordinary laws of enzyme kinetics are inadequate for providing the degree of control that is necessary for metabolism. Instead, many of the enzymes at control points display the property of responding with exceptional sensitivity to changes in metabolite concentrations. This property is commonly known as *co-operativity*, because it is thought to arise in many instances from 'co-operation' between the active sites of polymeric enzymes. (I shall consider more precise definitions of co-operativity later in this chapter, in Section 8.5, but for the present a precise definition is not required.) This chapter deals principally with examination of the main theories that have been proposed to account for co-operativity.

The interconversion of fructose 6-phosphate and fructose 1,6-bisphosphate illustrates another important aspect of metabolic control, namely that the immediate and ultimate products of a reac-tion are usually different. Although ATP is a substrate of the phospho-fructokinase reaction, the effect of glycolysis as a whole is to gener-ate ATP, in very large amounts if glycolysis is considered as the route into the tricarboxylate cycle and electron transport. Thus ATP must be regarded as a product of glycolysis, even though it is a substrate of the reaction at which glycolysis is controlled. Hence ordinary product inhibition of phosphofructokinase would produce the opposite effect from what is required for efficient control: to permit a steady supply of metabolic energy, phosphofructokinase ought to be inhibited by the ultimate product of the pathway, ATP, as in fact it is. This type of inhibition cannot be provided by the usual mechanisms, that is, by binding the inhibitor as a structural analogue of a substrate: in some cases these would bring about an unwanted effect; in others the ultimate product of a pathway might bear little structural resemblance to any of the reactants in the controlled step; for example, L-histidine bears little similarity to phosphoribosyl pyrophosphate, its biosynthetic precursor. To permit inhibition or activation by metabolically appropriate effectors, many controlled enzymes have evolved sites for effector binding that are separate from the catalytic sites. These are called *allosteric* sites, from the Greek for 'another solid', to emphasize the structural dissimilarity between substrate and effector, and enzymes that possess them are called allosteric enzymes.

Many allosteric enzymes are also co-operative, and vice versa, which is not surprising as both properties are important in metabolic control. This does not mean that the two terms are interchangeable, however: they describe two different properties and should be clearly distinguished. In many cases, the two properties were recognized separately: haemoglobin was known to be co-operative for over sixty years before the allosteric effect of 1,2-bisphosphoglycerate was described; the first enzyme in the biosynthesis of histidine has long been known to be allosteric, but has not been reported to be co-operative.

In concluding this introductory section, I should emphasize that metabolic control forms a large area of biochemical knowledge, and involves much more than it would be appropriate to include in a book about enzyme kinetics. To avoid the rather narrow view of control that this chapter may suggest, the reader should consult a more general treatment, such as that by Atkinson (1977).

8.2 Binding of oxygen to haemoglobin

Although haemoglobin is not an enzyme but a transport protein, study of binding of oxygen to it has contributed so much to an understanding of co-operativity that it would be inappropriate to discuss co-operativity without first discussing haemoglobin. Its co-operative properties were recognized by Bohr (1903) long before those of any enzyme, and much of the effort of developing theories of co-operativity was directed specifically at the co-operativity of haemoglobin.

The binding of oxygen to haemoglobin can be measured directly at equilibrium, so one does not have to rely on any questionable assumptions about the relationship between equilibrium binding and steady-state binding. Moreover, the existence of myoglobin, a non-co-operative analogue used for storing oxygen in muscle, permits a direct comparison that is not possible in other cases.

The binding of a ligand X to a simple monomeric protein E can be written as

$$E + X \xrightleftharpoons{K} EX$$

where K is the association constant, and the concentration of the complex at equilibrium is given by

$$[EX] = K[E][X] \tag{8.1}$$

Before we proceed further, it is necessary to draw attention to two differences between the symbols used in this chapter and those elsewhere in the book. First, equilibrium studies, particularly for haemoglobin, are usually discussed in terms of association constants rather than the dissociation constants that are more familiar to biochemists. This simplifies the appearance of many equations and, in any case, conversion of results from the literature to a different system would probably create more confusion than it would avoid. However, one important theory of co-operativity, the symmetry model of Monod, Wyman and Changeux (Section 8.7), is always discussed in terms of dissociation constants and will be discussed in these terms in this chapter. Second, abbreviated symbols for concentrations, such as x rather than $[X]$, will not be used in discussing co-operativity, because it is difficult to adapt this system to the concentrations of complicated species, such as EX_4, and because in equilibrium studies the total-protein concentration is often of the same order of magnitude as the total-ligand concentration. As a result, the free-ligand concentration may be much smaller than the total concentration, in contrast to the near equality that commonly obtains in steady-state kinetics.

We can define a quantity Y, known as the *fractional saturation*, as the fraction of binding sites that are occupied by ligand at any instant, i.e.

$$Y = \frac{\text{number of occupied binding sites}}{\text{total number of binding sites}} = \frac{[EX]}{[E] + [EX]}$$

Although the definition of Y refers to *numbers* of sites, whereas $[E]$ and $[EX]$ are *concentrations*, the two types of quantity are proportional to one another under defined conditions, and no inconsistency arises. Note that Y is a dimensionless number that must lie in the range 0–1. From equation 8.1, we obtain

$$Y = \frac{K[X]}{1 + K[X]} \tag{8.2}$$

This equation is the *Langmuir isotherm* (cf. Section 2.2), and describes a rectangular hyperbola through the origin when Y is plotted against $[X]$, which approaches a limit of $Y = 1$ when $[X]$ is large enough for 1 to be negligible compared with $K[X]$. Thus it closely resembles the Michaelis–Menten equation, with $1/K$ replacing K_m and 1 replacing V (by definition).

If the fractional saturation of myoglobin with oxygen is measured as a function of the partial pressure of oxygen (which may be regarded as analogous to the free concentration of oxygen in solution), the results do indeed obey equation 8.2, but the results for a similar experiment with haemoglobin fall on a different curve, which is strikingly *sigmoid* or S-shaped, as illustrated in *Figure 8.1*. Equation 8.2 cannot account for this curve and, from the time of Hill (1910) onwards, much effort has been devoted to the search for a plausible physical model that can.

There is an important physical difference between myoglobin and haemoglobin that sheds some light on the difference in properties between them. Myoglobin is a monomer, with a single polypeptide chain and a single oxygen-binding site per molecule, but haemoglobin

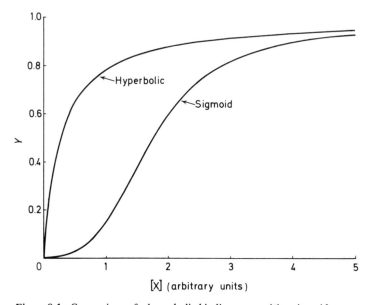

Figure 8.1 Comparison of a hyperbolic binding curve with a sigmoid curve

is a tetramer, consisting of four polypeptide chains, or *subunits*, per molecule, each with an oxygen-binding site. Although there are two distinct types of subunit in the haemoglobin molecule, two α-subunits and two β-subunits, they are similar in structure, not only to one another but also to myoglobin; to a first order of approximation, haemoglobin resembles a tetramer of myoglobin. It seems obvious with hindsight that the differences in binding properties between the two proteins are related to their different degrees of association, but it is worth noting that detailed information about the structures is rather recent (Kendrew *et al.*, 1960; Perutz *et al.*, 1960) and was not available to the earlier investigators of co-operativity.

8.3 Hill equation

Hill (1910) proposed the following equation, which is now commonly known as the *Hill equation*, to account for the oxygen-binding curves that he and others had observed for haemoglobin:

$$Y = \frac{K_h [X]^h}{1 + K_h [X]^h} \tag{8.3}$$

It is best to regard this equation as purely empirical and to refrain from attaching physical meaning to its parameters K_h and h. Hill himself wrote: 'I decided to try whether [equation 8.3] would satisfy the observations. My object was rather to see whether an equation *of this type* [italics in the original] would satisfy all the observations, than to base any direct physical meaning on [h and K_h].' This clear statement has unfortunately not discouraged later writers from giving spurious 'derivations' of equation 8.3, or from supposing h to have a simple physical meaning, such as the number n of ligand-binding sites on each molecule of protein. Such workers have sometimes been puzzled that h as experimentally determined is often non-integral and is rarely equal to n. However, as there is no reason why h should be an integer, it is in no way surprising if it is not. For reasons that will become clear in the next section, h cannot exceed n, at least in binding systems at equilibrium, and so it does provide a lower limit for n. Hill was able to fit all of the data at his disposal to equation 8.3, with values of h that ranged from 1 to 3.2, in no case approaching the actual number of oxygen-binding sites on each molecule of haemoglobin, which is now known to be 4.

If equation 8.3 is rearranged as follows:

$$Y/(1 - Y) = K_h [X]^h$$

$$\log [Y/(1 - Y)] = \log K_h + h \log [X] \tag{8.4}$$

it can be seen that a plot of $\log [Y/(1 - Y)]$ against $\log [X]$ should

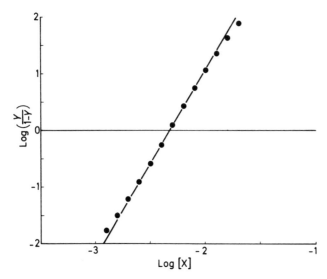

Figure 8.2 Hill plot. The line is drawn according to the Hill equation (equation 8.4), and does not fit the points exactly except in the middle of the range. It is often difficult to make measurements outside the range −1 to +1 on the ordinate, which corresponds to a range 0.09–0.91 in the value of Y

be a straight line of slope h. This plot, which is illustrated in *Figure 8.2*, is known as the *Hill plot* and provides a simple means of evaluating h and K_h. It has been found to fit a wide variety of binding data remarkably well for values of Y in the range 0.1–0.9, but deviations always occur at the extremes (as indicated in *Figure 8.2*) because equation 8.3 is at best only an approximation to a more complex relationship.

The exponent h is now commonly known as the *Hill coefficient.* It is widely used as an index of co-operativity, the degree of co-operativity being considered to increase as h increases. Taketa and Pogell (1965) have suggested a different parameter, the *co-operativity index*, R_x, defined as the ratio of the [X] values that give $Y = 0.9$ and $Y = 0.1$. Thus R_x has a more obvious experimental meaning than h, and is more convenient for discussing the properties of co-operative proteins in relation to their physiological roles. R_x has the further advantage that it is a purely empirical measure, not derived from any theoretical model of dubious validity. The relationship between the two indexes can be obtained by substituting $Y = 0.1$ and $Y = 0.9$ into equation 8.3 and solving for [X] in each case. For $Y = 0.1$, we have

$$0.1 = K_h[X]^h/(1 + K_h[X]^h)$$

Therefore

$$0.1 + 0.1K_h[X]^h = K_h[X]^h$$

Hence

$$[X] = \left(\frac{K_h}{9}\right)^{1/h}$$

Similarly, when $Y = 0.9$, we have $[X] = (9K_h)^{1/h}$. Dividing this second result by the first gives

$$R_X = 81^{1/h}$$

This expression is only as accurate as equation 8.3, of course, but

TABLE 8.1 Relationship between the two indexes of co-operativity

The table shows the relationship between the Hill coefficient, h, and the co-operativity index, R_X. The values are calculated on the assumption that the Hill equation (equation 8.3) holds exactly. Values of h greater than 1, or R_X less than 81, indicate *positive co-operativity*; values of h less than 1, or R_X greater than 81, indicate *negative co-operativity*.

h	R_X	h	R_X	h	R_X
0.5	6 560	1.5	18.7	5.0	2.41
0.6	1 520	2.0	9.00	6.0	2.08
0.7	533	2.5	5.80	8.0	1.73
0.8	243	3.0	4.33	10.0	1.55
0.9	132	3.5	3.51	15.0	1.34
1.0	81.0	4.0	3.00	20.0	1.25

that is adequate for most purposes. Some values calculated from it are shown in *Table 8.1*.

8.4 Adair equation

Adair (1925a, b), after determining that the molecular weight of haemoglobin was about four times as great as had previously been thought, suggested that there were four oxygen-binding sites per molecule, and that these sites were filled in a four-step process, as follows:

$$E + X \underset{}{\overset{4K_1}{\rightleftarrows}} EX$$

$$EX + X \xrightarrow{\frac{3}{2}K_2} EX_2$$

$$EX_2 + X \underset{}{\overset{\frac{2}{3}K_3}{\rightleftarrows}} EX_3$$

$$EX_3 + X \underset{}{\overset{\frac{1}{4}K_4}{\rightleftarrows}} EX_4$$

where the association constants K_1, K_2, K_3 and K_4 are so-called *intrinsic* constants. The 'statistical' factors $4, \frac{3}{2}, \frac{2}{3}$ and $\frac{1}{4}$ are written explicitly to permit a simple and direct correspondence between statements about the type of co-operativity and statements about the relationships between the K values. For example, if there are four identical binding sites that act independently, the four intrinsic constants are equal, i.e. $K_1 = K_2 = K_3 = K_4$. There is thus no co-operativity in this case. If each binding step facilitates the next one, we may say that there is positive co-operativity at each stage of the binding process, and $K_1 < K_2 < K_3 < K_4$. Similarly, if there is negative co-operativity at each stage, then $K_1 > K_2 > K_3 > K_4$. More complex relationships are also possible, as I shall discuss in the next section.

In Adair's model, the concentrations of the various species follow from the definitions of the association constants, thus:

$$[EX] = 4K_1 [E][X]$$

$$[EX_2] = \tfrac{3}{2}K_2 [EX][X] = 6K_1 K_2 [E][X]^2$$

$$[EX_3] = \tfrac{2}{3}K_3 [EX_2][X] = 4K_1 K_2 K_3 [E][X]^3$$

$$[EX_4] = \tfrac{1}{4}K_4 [EX_3][X] = K_1 K_2 K_3 K_4 [E][X]^4$$

These may be substituted into the definition of the fractional saturation to provide the expression

$$Y = \frac{\text{number of occupied sites}}{\text{total number of sites}}$$

$$= \frac{[EX] + 2[EX_2] + 3[EX_3] + 4[EX_4]}{4([E] + [EX] + [EX_2] + [EX_3] + [EX_4])}$$

$$= \frac{K_1 [X] + 3K_1 K_2 [X]^2 + 3K_1 K_2 K_3 [X]^3 + K_1 K_2 K_3 K_4 [X]^4}{1 + 4K_1 [X] + 6K_1 K_2 [X]^2 + 4K_1 K_2 K_3 [X]^3 + K_1 K_2 K_3 K_4 [X]^4}$$

$$(8.5)$$

This is known as the *Adair equation* for four sites. Similar equations can be derived in the same way for any number of sites. For example, the Adair equation for two sites takes the form

$$Y = \frac{K_1 [X] + K_1 K_2 [X]^2}{1 + 2K_1 [X] + K_1 K_2 [X]^2} \qquad (8.6)$$

In the general case of n sites, the numerator has n terms with the binomial coefficients for $(n - 1)$ as its coefficients, and the denominator has $(n + 1)$ terms with the binomial coefficients for n as its coefficients.

If all of the intrinsic association constants are equal, the Adair equation simplifies to the expression for a rectangular hyperbola through the origin. For example, equation 8.5 simplifies to

$$Y = \frac{K_1 [X] (1 + K_1 [X])^3}{(1 + K_1 [X])^4} = \frac{K_1 [X]}{1 + K_1 [X]} \tag{8.7}$$

The binding curve for haemoglobin is not a hyperbola, however, and it cannot therefore have four equal association constants.

If K_4 is sufficiently large compared with K_1, K_2 and K_3, equation 8.5 simplifies to

$$Y = \frac{K_1 K_2 K_3 K_4 [X]^4}{1 + K_1 K_2 K_3 K_4 [X]^4} \tag{8.8}$$

i.e. the Hill equation with $K_h = K_1 K_2 K_3 K_4$ and $h = 4$. However, there is no way in which equation 8.5 can be simplified to yield a Hill coefficient greater than 4. Moreover, the simplification expressed by equation 8.8 can at best describe only part of the saturation curve. If [X] is made sufficiently small, then, whatever the values of the association constants, the higher-order terms in equation 8.5 must eventually become smaller than $K_1 [X]$, and so at low [X] it must simplify not to equation 8.8 but to

$$Y = \frac{K_1 [X]}{1 + 4K_1 [X]}$$

Thus h must approach unity as [X] approaches zero; similarly h must approach unity as [X] approaches infinity. In general, for any values of the association constants, the Hill coefficient must approach unity both at very high and at very low ligand concentrations, and cannot exceed the number of binding sites at any ligand concentration.

Adair's model is the most general possible for a ligand binding to a pure non-associating protein at equilibrium. Even if the protein in any of its complexes exists as a set of equilibrating isomers, the *form* of the equation is unchanged, though the definitions of the parameters become more complicated. Only if the protein exists as a mixture of species that do not equilibrate does the binding equation become more complex. This case is referred to in Problem 8.2 at the end of this chapter, but is largely outside the scope of the discussion; from the practical point of view the main requirement is to ensure that experiments are done if possible with a pure protein, or, if pure protein cannot be obtained, then with a preparation which contains only one component that binds the ligand being studied. If this requirement is not satisfied, the observations may incorrectly suggest the occurrence of *negative* co-operativity; the appearance of positive co-operativity cannot be generated in this way.

8.5   Definition of co-operativity

The alert reader will have noticed that I have used the term *co-operativity* in two ways that are not precisely equivalent. It is appropriate, therefore, to examine the definition of this term more rigorously. When the nature of the binding process is precisely defined, as in Adair's model, it is convenient to consider whether each binding step facilitates or hinders the next one. For example, if one found with a four-site Adair equation that $K_1 < K_2 < K_3 > K_4$, one might reasonably describe this as a mixture of positive and negative co-operativity, with positive co-operativity between the first and second steps, and between the second and third, and negative co-operativity between the third and fourth.

A practical difficulty with basing the definition of co-operativity exclusively on the Adair equation is that one often wishes to specify whether a system is co-operative without knowing the Adair constants, or whether indeed the Adair equation is applicable. The usual practice in these circumstances is to define co-operativity in terms of the Hill coefficient, that is, to say that co-operativity is positive, zero or negative according to the sign of $(h - 1)$. Whitehead (1978) has suggested how this practice may be rationalized. Consider the quantity $Q = Y/\{[X](1 - Y)\}$: this is a constant equal to the association constant if there is only one binding site, or if there are n identical independent sites (i.e. the Adair constants satisfy the relationship $K_1 = K_2 = \ldots = K_n$). If, however, Q increases as $[X]$ increases, i.e. $dQ/d[X]$ is positive, it is clear that the binding is getting progressively stronger as more ligand binds: it is reasonable then to describe the system as positively co-operative at the particular value of $[X]$ at which $dQ/d[X]$ has been evaluated. Zero and negative co-operativity may be defined similarly. For any binding function, the sign of $dQ/d[X]$ is the same as that of $(h - 1)$, at any value of $[X]$; consequently, a definition of co-operativity in terms of the Hill plot is exactly equivalent to the more rational definition suggested by Whitehead. This conclusion is entirely independent of any consideration of whether the Hill equation has any physical or descriptive validity.

The definition of co-operativity in terms of the Hill coefficient is not necessarily equivalent to a definition in terms of Adair constants, and it remains therefore to consider what is the relationship between the two. Cornish-Bowden and Koshland (1975) have explored this question at a simple descriptive level, and have found that there is a fair but not exact correspondence between the two definitions. As an example, consider the data shown in *Figure 8.3*. The curve has a slope greater than 1 at low ligand concentrations, is equal to 1 close to half-saturation, and is less than 1 at high ligand concentrations.

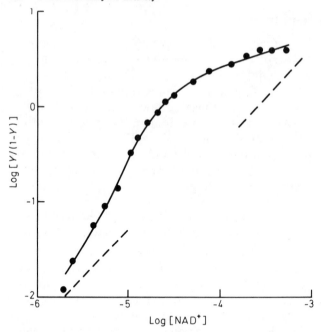

Figure 8.3 Hill plot for the binding of NAD$^+$ to yeast glyceraldehyde 3-phosphate dehydrogenase. The plot shows data of Cook and Koshland (1970) recalculated as described by Cornish-Bowden and Koshland (1975). The shape of the curve suggests that the Adair constants satisfy the relationship $K_1 < K_2 \simeq K_3 > K_4$, in agreement with the following values found by curve-fitting (Cornish-Bowden and Koshland, 1970): $K_1 = 4.6 \times 10^3$ M^{-1}, $K_2 = 1.4 \times 10^5$ M^{-1}, $K_3 = 7.5 \times 10^4$ M^{-1}, $K_4 = 3.5 \times 10^3$ M^{-1}. The figure is reproduced with permission from *J. mol. Biol.* (1975) **95**, 201–212. Copyright © by Academic Press Inc. (London) Ltd

This suggests that the Adair constants might obey a relationship of the form $K_1 < K_2 \simeq K_3 > K_4$, and this was indeed found when the data were fitted to the Adair equation. Cornish-Bowden and Koshland examined many calculated Hill plots and found that this sort of correspondence applied in most cases. Thus although definitions of co-operativity based on the Hill plot and the Adair equation are not equivalent, they are qualitatively similar and no great harm will follow from continuing to use both as appropriate: the Hill-plot definition applies more generally, but the Adair-equation definition has greater physical meaning in circumstances where it can be used.

In kinetic experiments it is not usually possible to measure Y directly. Instead, its value is inferred from the kinetic observations, by assuming that the rate of reaction v is proportional to the extent of saturation of the enzyme, that is, to assume that $Y = v/V^{app}$, where V^{app} is the extrapolated rate at saturation, other conditions being kept constant. This assumption is unlikely to be true in all cases, and may indeed be true rather rarely, because it implies, in a co-operative enzyme, that interactions which alter the binding properties of the enzyme have no effect on its rate constants for catalysis. Even if this first assumption is valid, detailed analysis normally requires a further assumption that binding in the steady state corresponds to binding at equilibrium, which is of course the same assumption that is widely recognized to be hazardous in simple systems (*see* Section 2.3). So although it is often useful to assume that $Y = v/V^{app}$ and then treat the resulting Y values as if they were measured at equilibrium, one should not forget that this is a first approximation that may need to be discarded in the light of additional information. With this convention, kinetic data are often described in the same terms as those developed for equilibria. The Hill plot, for example, can still be used for defining co-operativity: in the kinetic case it becomes a plot of log $[v/(V^{app} - v)]$ against log $[S]$, where S is the variable substrate. It should be noted that a value of V^{app} is required before a Hill plot can be made, but this can usually be obtained by extrapolation.

8.6 Induced fit

Early theories of haemoglobin co-operativity assumed that the oxygen-binding sites on each molecule of haemoglobin would have to be close enough together to interact electronically. This assumption was made explicit by Pauling (1935), but was already implied in Hill's and Adair's ideas. When the three-dimensional structure of haemoglobin was determined (Perutz *et al.*, 1960), however, the haem groups proved to be too far apart (2.5–4.0 nm) to interact in any of the ways that had been envisaged. Nonetheless, long-range interactions do occur, in other proteins as well as haemoglobin, and all modern theories account for these in terms of *protein flexibility*. In that limited sense they derive from the *theory of induced fit* of Koshland (1958, 1959a, b), and the purpose of this section is to examine the experimental and theoretical basis of this theory.

The high degree of specificity that enzymes display towards their substrates has impressed biochemists since the earliest studies of enzymes, even before anything was known about their physical and chemical structures. Fischer (1894) was particularly impressed by the ability of living organisms to discriminate totally between sugars that differed only slightly and at atoms remote from the sites of reaction. To explain this ability, he proposed that the active site of an enzyme was a negative imprint of its substrate(s), and that it would catalyse the reactions only of compounds that fitted precisely. This is similar to the mode of action of an ordinary (non-Yale) key in a lock, and the theory is known as Fischer's *lock-and-key model* of enzyme action. For many years, it seemed to explain all of the known facts of enzyme specificity, but as more detailed research was carried out there were numerous observations that were difficult to account for in terms of a rigid active site of the type that Fischer had envisaged. For example, the occurrence of enzymes for two-substrate reactions that require the substrates to bind in the correct order provides one type of evidence, as mentioned in Section 6.2. A more striking example, noted by Koshland, was the failure of water to react in several enzyme-catalysed reactions where the lock-and-key model would lead one to expect it to react. Consider, for example, the reaction catalysed by hexokinase:

$$\text{glucose} + \text{MgATP}^{2-} \rightleftharpoons \text{glucose 6-phosphate} + \text{MgADP}^{-}$$

The enzyme from yeast is not particularly specific for its sugar substrate: it will accept not only glucose but other sugars, such as fructose and mannose. Water does not react, however, even though it can scarcely fail to saturate the active site of the enzyme, at a concentration of 56 M, about 7×10^6 times the Michaelis constant for glucose, and chemically it is at least as reactive a compound as the sugars that do react.

Koshland argued that these and other observations provided strong evidence for a *flexible* active site; he proposed that the active site of an enzyme has the *potential* to fit the substrate precisely, but that it does not adopt the negative substrate form until the substrate binds. This conformational change accompanying substrate binding brings about the proper alignment of the catalytic groups of the enzyme with the site of reaction in the substrate. With this hypothesis, the properties of yeast hexokinase can easily be explained: water can certainly bind to the active site of the enzyme, but it lacks the bulk to force the conformational change necessary for catalysis.

Koshland's theory is known as the induced-fit hypothesis, to emphasize its differences from Fischer's theory, which assumes that the fit between enzyme and substrate pre-exists and does not need to be induced. The lock-and-key analogy can be pursued a little further by likening Koshland's conception to a Yale lock, in which the key must not merely fit but must also realign the tumblers before it will turn.

The induced-fit theory has had important consequences in several branches of enzymology, but it was particularly important in the understanding of allosteric and co-operative phenomena in proteins, because it provided a simple and plausible explanation of long-range interactions. Provided that a protein combines rigidity with flexibility in a controlled and purposive way, like a pair of scissors, a substrate-induced conformational change at one point in the molecule may be communicated over several nanometres to any other point.

8.7 Symmetry model of Monod, Wyman and Changeux

Both co-operative interactions in haemoglobin and allosteric effects in many enzymes require interactions between sites that are widely separated in space. A striking example of this requirement is provided by the allosteric inhibition of phosphoribosyl-ATP pyrophosphorylase by histidine: Martin (1963) found that mild treatment of the enzyme with mercury(II) ions destroyed the sensitivity of the catalytic activity to histidine but did not affect either the uninhibited activity or the binding of histidine. In other words, the metal ion interfered neither with the catalytic site nor with the allosteric site, but with the connection between them. Monod, Changeux and Jacob (1963) studied many examples of co-operative and allosteric phenomena, and concluded that they were closely related and that conformational flexibility probably accounted for both. Subsequently, Monod, Wyman and Changeux (1965) proposed a general model to explain both phenomena within a simple set of postulates. The model is often referred to as the *allosteric model*, but the term *symmetry model* is preferable because it emphasizes the principal difference between it and later models and because it avoids the contentious association between allosteric and co-operative phenomena.

The symmetry model starts from the observation that most co-operative proteins contain several subunits in each molecule. Indeed, this must be so for binding co-operativity at equilibrium though it is not required in kinetic co-operativity, as I shall discuss in Section 8.9. For simplicity I shall describe the symmetry model in terms of a tetrameric protein, but any number of subunits greater than one is possible. The model includes the following postulates:

(1) Each subunit can exist in two different conformations, designated R and T. These labels originally stood for *relaxed* and *tense*, respectively, but they are nowadays commonly regarded simply as labels.

(2) All subunits of a molecule must occupy the same conformation at any time; hence, for a tetrameric protein, the conformational states R_4 and T_4 are the only two permitted, conformational mixtures such as $R_3 T$ being forbidden.

(3) The two states of the protein are in equilibrium, with an equilibrium constant $L = [T_4]/[R_4]$.

(4) A ligand can bind to a subunit in either conformation, but the dissociation constants are different: $K_R = [R][X]/[RX]$ for each R subunit, $K_T = [T][X]/[TX]$ for each T subunit, and $K_R/K_T = c$ by definition.

These postulates imply the following set of equilibria between the various forms of the protein:

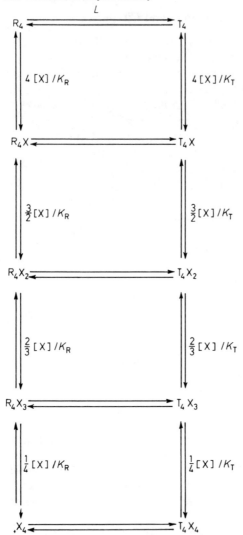

The concentrations of the ten forms of the protein are related by the following expressions:

$$[R_4 X] = 4[R_4][X]/K_R$$

$$[R_4 X_2] = \tfrac{3}{2}[R_4 X][X]/K_R = 6[R_4][X]^2/K_R^2$$

$$[R_4 X_3] = \tfrac{2}{3}[R_4 X_2][X]/K_R = 4[R_4][X]^3/K_R^3$$

$$[R_4 X_4] = \tfrac{1}{4}[R_4 X_3][X]/K_R = [R_4][X]^4/K_R^4$$

$$[T_4] = L[R_4]$$

$$[T_4 X] = 4[T_4][X]/K_T = 4Lc[R_4][X]/K_R$$

$$[T_4 X_2] = 6[T_4][X]^2/K_T^2 = 6Lc^2[R_4][X]^2/K_R^2$$

$$[T_4 X_3] = 4[T_4][X]^3/K_T^3 = 4Lc^3[R_4][X]^3/K_R^3$$

$$[T_4 X_4] = [T_4][X]^4/K_T^4 = Lc^4[R_4][X]^4/K_R^4$$

In each equation, the 'statistical' factor 4, $\frac{3}{2}$, etc. results from the fact that the dissociation constants are defined in terms of individual sites but the expressions are written for complete molecules. For example, $K_R = [R][X]/[RX] = \frac{3}{2}[R_4 X][X]/[R_4 X_2]$, because there are three unliganded R subunits in each $R_4 X$ molecule and two liganded R subunits in each $R_4 X_2$ molecule. The fractional saturation Y is defined as before and is given by the equation

$$Y =$$

$$\frac{[R_4 X] + 2[R_4 X_2] + 3[R_4 X_3] + 4[R_4 X_4] + [T_4 X] + 2[T_4 X_2] + 3[T_4 X_3] + 4[T_4 X_4]}{4([R_4] + [R_4 X] + [R_4 X_2] + [R_4 X_3] + [R_4 X_4] + [T_4] + [T_4 X] + [T_4 X_2] + [T_4 X_3] + [T_4 X_4])}$$

$$= \frac{(1 + [X]/K_R)^3 [X]/K_R + Lc(1 + c[X]/K_R)^3 [X]/K_R}{(1 + [X]/K_R)^4 + L(1 + c[X]/K_R)^4} \qquad (8.9)$$

The shape of the saturation curve defined by this equation depends on the values of L and c, as may be illustrated by assigning some extreme values to these constants. If $L = 0$, which means that the T form of the protein does not exist under any conditions, equation 8.9 simplifies to $Y = [X]/(K_R + [X])$, because the factor $(1 + [X]/K_R)^3$ can be cancelled from numerator and denominator in this case. This simplified equation is the Langmuir isotherm, which predicts hyperbolic binding. A similar simplification occurs if L approaches infinity: in this case, $Y = [X]/(K_T + [X])$. It follows that deviations from hyperbolic binding can occur with this model only if both conformational states of the protein exist in significant amounts. This is reasonable, because if there is only one form of the protein the model is the same as Adair's model with independent and identical binding sites (cf. equation 8.7).

Hyperbolic binding also arises in the symmetry model in a third special case: if $c = 1$, it is again possible to cancel the common factor $(1 + [X]/K_R)^3$, leaving a Langmuir-isotherm expression. This illustrates the reasonable conclusion that if the ligand binds equally well to the two states of the protein, the relative amounts of them are irrelevant to the binding behaviour. Apart from these special cases, equation 8.9 predicts positively co-operative binding, though this may not be obvious unless we consider the limiting case of $c = 0$, in which X binds *only* to the R state. Although this limiting case is a natural application of the idea of induced fit, it is not an essential characteristic of the symmetry model as proposed by Monod, Wyman and Changeux. When $c = 0$, equation 8.9 simplifies to

$$Y = \frac{(1 + [X]/K_R)^3 \, [X]/K_R}{L + (1 + [X]/K_R)^4} \qquad (8.10)$$

At very high values of $[X]$, when $[X]/K_R \gg L$, the term L in the denominator is negligible and the expression as a whole factorizes to the Langmuir isotherm; at low values of $[X]$, however, the term L dominates the denominator, so the saturation curve must rise very slowly from the origin. In other words, the curve must be sigmoid if L is large compared with 1.

To see why the more general expression, equation 8.9, predicts co-operativity, we must examine its relationship with the Adair equation. If the terms $(1 + [X]/K_R)^3$, etc. in equation 8.9 are multiplied out and rearranged, the equation assumes the form of the Adair equation for four sites (equation 8.5), with the four Adair association constants defined as

$$K_1 = \frac{1 + Lc}{K_R(1 + L)}$$

$$K_2 = \frac{1 + Lc^2}{K_R(1 + Lc)}$$

$$K_3 = \frac{1 + Lc^3}{K_R(1 + Lc^2)}$$

$$K_4 = \frac{1 + Lc^4}{K_R(1 + Lc^3)}$$

If we now examine the ratio of any pair of Adair constants, for example K_3/K_2, we find

$$\frac{K_3}{K_2} = \frac{(1 + Lc^3)(1 + Lc)}{(1 + Lc^2)^2} = \frac{1 + Lc(c^2 + 1) + L^2c^4}{1 + 2Lc^2 + L^2c^4}$$

As $Lc(c^2 + 1) \geqslant 2Lc^2$ for all positive values of L and c, it follows that the right-hand fraction cannot be less than 1, i.e. $K_3 \geqslant K_2$. Similar results apply to all other pairs of constants, and also apply if the model is generalized to include more than four binding sites and more than two conformations. The symmetry model must therefore give rise to positive co-operativity and cannot give rise to negative co-operativity. Some representative binding curves calculated from equation 8.9 are shown in *Figure 8.4*.

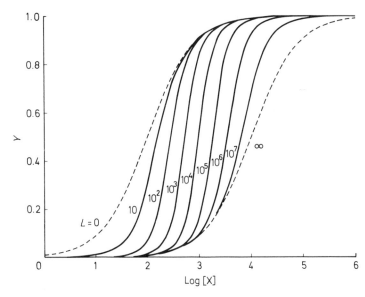

Figure 8.4   Binding curves for the symmetry model (equation 8.9), with $c = 0.01$ and $L = 0-\infty$ as indicated. Arbitrary units are used for [X]. The curve for $L = 0$ is the binding curve for the pure R state, and the curve for $L = \infty$ is the binding curve for the pure T state. Both of these extreme curves would be hyperbolic (cf. *Figure 8.1*) if Y were plotted against [X] rather than log [X], whereas the intermediate curves would be sigmoid

Monod, Wyman and Changeux distinguished between *homotropic* effects, or interactions between identical ligands, and *heterotropic* effects, or interactions between different ligands, such as a substrate and an allosteric effector. Although the symmetry model requires homotropic effects to be positively co-operative, it places no corresponding restriction on heterotropic effects, and can account for these with no extra complexity; this is indeed one of its most satisfying features. If ligand A binds preferentially to the R state of the protein, that is, the state to which X binds preferentially, but at a different site from X, it will clearly facilitate binding of X by increasing the availability of molecules in the R state; it will thus act as a positive heterotropic effector, or allosteric activator. Conversely, if I binds preferentially to the T state, which binds X weakly or not at all, it will hinder the binding of X by decreasing the availability of molecules in the R state; it will thus act as a negative heterotropic effector or allosteric inhibitor. If all binding is exclusive, rather than preferential, i.e. each ligand binds either to the R or to the T state but not to both, the resulting binding equation for X is particularly simple, and is examined in Problem 8.4 at the end of this chapter.

Several enzymes do in fact behave as the symmetry model predicts, for example phosphofructokinase from *Escherichia coli*, which was

studied by Blangy, Buc and Monod (1968). Over a wide range of concentrations of ADP, an allosteric activator, and phospho-*enol*-
pyruvate, an allosteric inhibitor, the binding of one substrate, fructose 6-phosphate, is closely in accordance with the model. Nonetheless, the symmetry model cannot be regarded as the complete explanation of binding co-operativity, because there are some observed
phenomena that it cannot explain, such as negative co-operativity,
and some of its postulates are not altogether convincing. The central
assumption of conformational symmetry is not readily explainable
in structural terms, for example. Moreover, for many enzymes it is
necessary to postulate the occurrence of a *perfect K system*, which
means that the R and T states of the enzyme have identical catalytic
properties despite having grossly different binding properties. These
and other questionable aspects of the symmetry model have stimu-
lated the search for alternatives.

8.8 Sequential model of Koshland, Némethy and Filmer

Although the symmetry model incorporates the idea of purposive
conformational flexibility, it departs from the induced-fit theory in
permitting ligands to bind to both R and T conformations, albeit
with different binding constants. Koshland, Némethy and Filmer
(1966) showed that a more orthodox application of induced fit could
account for co-operativity equally well. Like Monod, Wyman and
Changeux, they postulated two conformations, which they termed
the A and B conformations (corresponding to the T and R conformations, respectively), but they assumed that the B conformation was
induced by ligand binding so that X binds only to the B conformation and the B conformation exists only with X bound to it.

    Koshland, Némethy and Filmer assumed that co-operativity arose
because the properties of each subunit were modified by the conformational states of the neighbouring subunits. This assumption is implicit in the symmetry model, but is emphasized in the sequential
model, which is much more concerned with the details of interaction, and avoids the arbitrary assumption that all subunits must
exist simultaneously in the same conformation. Hence conformational
hybrids, such as AB_3, $A_2 B_2$, etc., are not merely allowed, but are
required by the assumption of strict induced fit.

    Because the symmetry model was not concerned with the details
of subunit interactions, there was no need in the previous section to
consider the geometry of subunit association, i.e. the quaternary
structure of the protein. In contrast, the sequential model does
require the geometry to be considered, because different arrange-
ments of subunits result in different binding equations. The emphasis
on geometry and the need to treat each geometry separately have

given rise to the widespread but erroneous idea that the sequential model is more general and more complicated than the symmetry model; but for any given geometry the two models are about equally complex and neither is a special case of the other. Both models can be generalized into the same general model (Haber and Koshland, 1967), by relaxing the symmetry requirement of the symmetry model and the strict induced-fit requirement of the sequential model, but it is questionable whether this is worth while, because the resulting equation is too complicated to use. In some contexts, it is helpful to refer to the ordinary form of the sequential model as the *simplest* sequential model, to distinguish it from the general model.

In explaining the characteristics of the sequential model, I shall use the *square* geometry as an illustration: in this case the subunits of a tetramer are assumed to be arranged so that each can interact with two neighbours, but not with the third. *Tetrahedral* and *linear* geometries are also possible for four subunits, but the method of analysis is the same as for the square case, though the results differ in detail. If the A conformation is drawn as a circle, and the B conformation as a square, the six possible species for the binding of X are as follows:

$$A_4 \qquad A_3 BX \qquad A_2 B_2 X_2 \qquad AB_3 X_3 \qquad B_4 X_4$$

Note that there are two ways of drawing $A_2 B_2 X_2$, which must be considered separately, because the subunit contacts are different. The concentration of each species can be expressed by considering the various changes needed to obtain it from the standard state, which is taken as the unliganded protein, A_4. For example, to obtain $AB_3 X_3$ from A_4, the following changes must occur:

(1) Three subunits must undergo the conformational change
       A → B. This is represented by K_t^3, where K_t is the notional
       equilibrium constant [B]/[A] for an isolated subunit. In the
       simplest sequential model, K_t is tacitly assumed to be small,
       in keeping with the assumption that the B conformation
       occurs only when induced by the binding of X.

(2) Three molecules of X must bind to three B subunits. This is
       represented by $K_x^3 [X]^3$, where K_x is the association constant
       [BX]/[B][X] for binding of X to an isolated B subunit.

(3) In the square geometry there are four interfaces between

neighbouring subunits. In the standard state, A_4, each interface can be designated AA, as both touching subunits are in the A conformation. In $AB_3 X_3$ there are no AA interfaces, however; instead, there are two AB interfaces and two BB interfaces. The appearance of these is allowed for by two kinds of *subunit-interaction terms* K_{AB} and K_{BB}: K_{AB} represents the equilibrium constant $[AB]/[AA]$ for the conversion of an AA interface into an AB interface, and similarly K_{BB} is defined as $[BB]/[AA]$. The complete change from A_4 to $AB_3 X_3$ requires $K_{AB}^2 K_{BB}^2$. The constant K_{AB} can also be regarded as an *absolute* measure of the stability of the AB interface, but then another quantity, K_{AA}, is required for the AA interface, which is arbitrarily assigned a value of unity. It is simpler and just as rigorous to regard K_{AB} as a *relative* measure of the stability of the AB interface compared with that of the AA interface, as in the definition given above, and then no extra constant K_{AA} is needed. Similarly, it is simplest to regard K_{BB} as a measure of the stability of the BB interface compared with the AA interface.

(4)   Finally, a statistical factor of 4 is required, because there are four equivalent ways of choosing three out of four subunits. The word 'equivalent' is necessary here, because non-equivalent choices must be treated separately: for $A_2 B_2 X_2$ the diagonal arrangement has a statistical factor of 2 and the contiguous arrangement has one of 4.

All of these terms may now be multiplied together to give an expression for the concentration of $AB_3 X_3$:

$$[AB_3 X_3] \ = \ 4[A_4] K_x^3 K_t^3 K_{AB}^2 K_{BB}^2 [X]^3$$

and the corresponding expressions for the other species may be derived in a similar way:

$$[A_3 BX] \ = \ 4[A_4] K_x K_t K_{AB}^2 [X]$$

$$[A_2 B_2 X_2] \ = \ 2[A_4] K_x^2 K_t^2 (2K_{AB}^2 K_{BB} + K_{AB}^4)[X]^2$$

$$[B_4 X_4] \ = \ [A_4] K_x^4 K_t^4 K_{BB}^4 [X]^4$$

These four equations may be combined into the following expression for the fractional saturation:

$$Y \ = \ \frac{[A_3 BX] + 2[A_2 B_2 X_2] + 3[AB_3 X_3] + 4[B_4 X_4]}{4([A_4] + [A_3 BX] + [A_2 B_2 X_2] + [AB_3 X_3] + [B_4 X_4])}$$

$$= \ \frac{K_x K_t K_{AB}^2 [X] + K_x^2 K_t^2 (2K_{AB}^2 K_{BB} + K_{AB}^4)[X]^2 + 3K_x^3 K_t^3 K_{AB}^2 K_{BB}^2 [X]^3 + K_x^4 K_t^4 K_{BB}^4 [X]^4}{1 + 4K_x K_t K_{AB}^2 [X] + 2K_x^2 K_t^2 (2K_{AB}^2 K_{BB} + K_{AB}^4)[X]^2 + 4K_x^3 K_t^3 K_{AB}^2 K_{BB}^2 [X]^3 + K_x^4 K_t^4 K_{BB}^4 [X]^4}$$

$$(8.11)$$

As written this equation appears rather complicated, because it allows separately for every aspect of ligand binding. It is less complicated than it appears, however, because some of the constants appear always in the same combinations; for example, the assumption of strict induced fit means that the change in conformation and binding of ligand occur together or not at all, so $K_x K_t [X]$ always occurs as a product. Less obvious combinations also occur repeatedly because the subunit interactions are not independent of ligand binding; for example, K_{BB} cannot occur in a product that does not contain $K_x^2 K_t^2 [X]^2$, because a BB interaction is possible only if there are two B subunits. By writing $\bar{K} = K_x K_t K_{BB}$ and $c^2 = K_{AB}^2 / K_{BB}$, we can simplify equation 8.11 considerably, as follows:

$$Y = \frac{c^2 \bar{K}[X] + c^2 (2 + c^2)\bar{K}^2 [X]^2 + 3c^2 \bar{K}^3 [X]^3 + \bar{K}^4 [X]^4}{1 + 4c^2 \bar{K}[X] + 2c^2 (2 + c^2)\bar{K}^2 [X]^2 + 4c^2 \bar{K}^3 [X]^3 + \bar{K}^4 [X]^4} \qquad (8.12)$$

The meanings of the two new parameters $\bar{K}$ and c are as follows: $\bar{K}$ is the geometric mean of the four Adair constants, because $\bar{K}^4 = K_x^4 K_t^4 K_{BB}^4 = [B_4 X_4]/[A_4][X]^4$ is the association constant for the complete four-step binding; and c is a measure of the stability of the AB interaction compared with the AA and BB interactions. Equation 8.12 is not only simpler than equation 8.11, it is also the form required for fitting the sequential model to experimental results; this is not possible with equation 8.11, because it is 'overdetermined' and any change in K_x, for example, can be exactly compensated for by an opposite change in K_t or K_{BB}. In other words, it is possible in principle to determine $\bar{K}$ and c from measurements of Y as a function of $[X]$, but it is not possible to determine the separate values of K_x, K_t, K_{AB} and K_{BB} from such measurements. The effects that $\bar{K}$ and c have on the binding curve may be judged from the representative curves shown in *Figure 8.5*.

   Comparison of equation 8.12 with equation 8.5 provides the definitions of the Adair constants as they apply to the sequential model:

$K_1 \;=\; c^2 \bar{K}$

$K_2 \;=\; \frac{1}{3}(2 + c^2)\bar{K}$

$K_3 \;=\; 3\bar{K}/(2 + c^2)$

$K_4 \;=\; \bar{K}/c^2$

These relationships allow a simple determination of the types of co-operativity permitted by the simplest sequential model. The ratio of any pair of Adair constants, e.g. K_3/K_2, depends only on the value of c:

$K_3/K_2 \;=\; 9/(2 + c^2)^2$

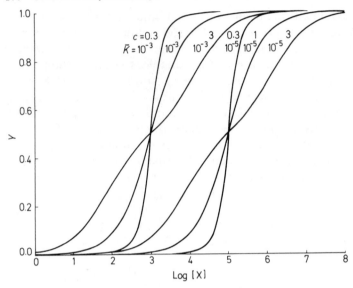

Figure 8.5 Binding curves for the sequential model (equation 8.12), with the values of $\bar{K}$ and c indicated. Arbitrary units are used for [X]. The location of the half-saturation point of each curve is determined solely by $\bar{K}$ and the shape is determined solely by c

Therefore, $K_3 < K_2$ if $c > 1$ and $K_3 > K_2$ if $c < 1$, and the same inequalities apply for other ratios of constants. The postulates of the sequential model allow the mixed interaction AB to be either weaker ($c < 1$) or stronger ($c > 1$) than the average of the pure interactions AA and BB, and consequently the model can account as easily for negative co-operativity ($K_1 > K_2 > K_3 > K_4$) as for positive co-operativity ($K_1 < K_2 < K_3 < K_4$), unlike the symmetry model, which is restricted to positive co-operativity. Mixed positive and negative co-operativity, as seen, for example, in *Figure 8.3* (p. 158), cannot be explained by the simplest versions of either model and requires additional, more complicated, assumptions.

Koshland, Némethy and Filmer (1966) showed that equation 8.12, and the corresponding equation derived assuming tetrahedral geometry, fitted the oxygen–haemoglobin saturation curve about as well as the equation for the symmetry model. It would therefore be difficult to distinguish between the models on the basis of the data for haemoglobin, or any other positively co-operative protein. This does not, however, mean that saturation curves can never permit a discrimination between models: for a protein showing negative co-operativity, the saturation curve alone provides sufficient evidence for ruling out the symmetry model. In fact, several examples of negative co-operativity are now known and, although some of these results may reflect impure samples of protein, the binding of NAD$^+$ to glyceraldehyde 3-phosphate dehydrogenase from rabbit muscle displays such an extreme degree of negative co-operativity (Conway and Koshland, 1968) that it cannot be explained by impurities.

Although the sequential model was originally proposed as a way of accounting for homotropic interactions, Kirtley and Koshland (1967) subsequently extended it to take account of heterotropic interactions also. Their treatment is not conceptually difficult and develops naturally from the theory for homotropic interactions, but it leads to much more complicated equations and requires many different possible sets of assumptions to be considered. For these reasons it has been very little used in the analysis of experimental studies, and I shall not discuss it further.

8.9 Other models for co-operativity at equilibrium

Frieden (1967) and Nichol, Jackson and Winzor (1967) independently suggested that co-operativity might in some circumstances result from the existence of an equilibrium between protein forms in different states of aggregation, such as a monomer A and a tetramer B_4 :

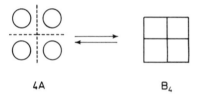

$4A$ B_4

If the two forms have different affinities for ligand, this model is conceptually rather similar to the symmetry model, and it predicts co-operativity for much the same reasons. The equations that describe it are more complicated, however, because the extent of co-operativity varies with the protein concentration. The reason for this is that the proportion of tetramer increases as the protein concentration increases; by contrast, the constant L in the symmetry model that defines the equilibrium between unliganded forms is independent of protein concentration.

The association–dissociation model is much more accessible to experimental verification than those I have considered previously. Not only should the degree of co-operativity vary with the protein concentration, but a large change in molecular weight should accompany ligand binding. Either of these effects should be easy to detect, and neither is predicted by the other models. Reversible association does appear to provide a complete explanation for the co-operativity of binding various nucleotides to glutamate dehydrogenase (Frieden and Colman, 1967). It may also contribute to the co-operativity of haemoglobin, which dissociates to a dimeric form at high salt concentrations, but it cannot explain haemoglobin co-operativity completely, because it is observed under many conditions where no dissociation takes place. For any protein, it is an obvious precaution to check whether the observed co-operativity varies with the protein concentration; if it does, then an association–dissociation model must be considered as a possible explanation.

8.10 Kinetic models of co-operativity

All of the models discussed in the earlier part of this chapter have been essentially equilibrium models; they can be applied to kinetic experiments only by assuming that v/V can be taken as a true measure of Y. However, co-operativity can also arise for purely kinetic reasons, in mechanisms that would show no co-operativity if binding could be measured at equilibrium. Although kinetic co-operativity can in principle occur in a reaction with only one substrate (Rabin,

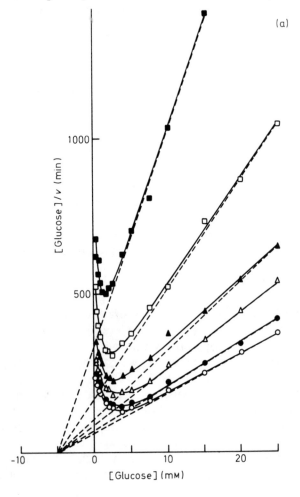

(a)

1967), there do not appear to be experimental examples, and in this
section I shall discuss kinetic co-operativity in two-substrate reactions.

Glucokinase is an enzyme found in the livers of vertebrates. It is
monomeric over a wide range of conditions, including those used in
its assay (Holroyde *et al.*, 1976; Cárdenas, Rabajille and Niemeyer,
1978), but it shows marked deviations from Michaelis–Menten kinetics when the glucose concentration is varied at constant concentrations of the other substrate, MgATP^{2-} (Storer and Cornish-Bowden,
1976b), as illustrated in *Figure 8.6*. When replotted as a series of Hill
plots, the data show *h* values ranging from about 1.5 at saturating
MgATP^{2-} to a low value, possibly 1.0, at vanishingly small MgATP^{2-}
concentrations. Glucokinase appears, therefore, to be a clear example
of a monomeric co-operative enzyme. Deviations in the opposite
sense are also possible: hexokinase type L$_I$ from wheat-germ is also
monomeric under assay conditions and displays negative co-operativity when glucose is the variable substrate (Meunier *et al.*, 1974). In

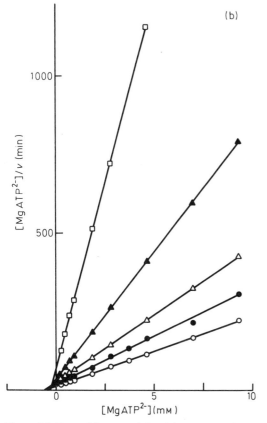

Figure 8.6 Data of Storer and Cornish-Bowden (1976) for the kinetics of glucokinase as
a function of (a) glucose (opposite) and (b) MgATP^{2-}. In each case the different symbols
represent different concentrations of the constant substrate

other respects its kinetic behaviour resembles that of glucokinase, and neither enzyme shows deviations from Michaelis–Menten kinetics with respect to the second substrate, $MgATP^{2-}$.

The existence of examples such as these of co-operative mono-meric enzymes requires serious attention to be given to kinetic models of co-operativity. I shall consider two in this section. The older is due to Ferdinand (1966), who pointed out that if a steady-state rate equation is derived for the random-order ternary-complex mechanism (Section 6.2) without assuming that substrates bind at equilibrium, the result is much more complicated than equation 6.3 (p. 107) and contains terms in the squares of both substrate concen-trations. Ferdinand suggested that a model of this kind, which he called a *preferred-order* mechanism, might provide an explanation of co-operativity. Although it is clear enough from consideration of the King–Altman method that deviations from Michaelis–Menten kinetics ought to occur with this mechanism, this explanation is rather abstract and algebraic and cannot easily be translated into concep-tually simple terms. The point is that both pathways for substrate binding must be assumed to make significant contributions to the total flux through the reaction, but the relative magnitudes of these contributions change as the substrate concentrations change. Thus the observed behaviour corresponds approximately to one pathway at low concentrations, but to the other at high concentrations.

Ricard, Meunier and Buc (1974) developed an alternative model of kinetic co-operativity from earlier ideas of Rabin (1967) and Whitehead (1970). Their model is known as a *mnemonical* model (from the Greek for memory), because it depends on the idea that the enzyme changes conformation relatively slowly, and is thus able to 'remember' the conformation that it had during a recent catalytic cycle. It is shown (in simplified form) in the following scheme:

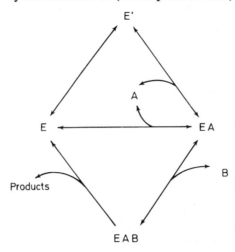

Its essential characteristics are as follows: it postulates that there are two forms E and E' of the free enzyme that differ in their affinities for A, the first substrate to bind; in addition, equilibration between E, E', A and EA must be slow relative to the maximum flux through the reaction. With these postulates, one can readily account for the sort of behaviour observed with glucokinase and wheat-germ hexo-kinase. As the concentration of B is lowered, the rate at which EA is converted into EAB and thence into products must eventually become slow enough for E, E', A and EA to equilibrate. So at van-ishing concentrations of B, the binding of A should behave as an ordinary equilibrium, with no co-operativity because there is only a single binding site. At high concentrations of B, on the other hand, it becomes possible for EA to be removed so fast that it cannot equilibrate and the laws of equilibria cease to apply. Deviations from Michaelis–Menten kinetics are then possible because at low concen-trations of A the two forms of free enzyme can equilibrate but at high concentrations they cannot.

Problems

8.1   Watari and Isogai (1976) have proposed a plot of $\log \{ Y/[X](1 - Y)\}$ against $\log [X]$ as an alternative to the Hill plot. What is the slope of this plot (expressed in terms of the Hill coefficient h)? What advantage does it have over the Hill plot?

8.2   Consider an equimolar mixture of two monomeric proteins that both bind the same ligand X according to the Langmuir isotherm, with different association constants K'_1 and K'_2. Show that the binding equation for the mixture of two proteins has the same form as the Adair equation for two sites on the same protein, and give the expressions for the Adair constants K_1 and K_2 (defined by equation 8.6) in terms of K'_1 and K'_2. What restric-tion does this model place on the allowed values of K_1 and K_2? Differentiate the expression for Y twice with respect to $[X]$ and show that the resulting second derivative is negative at all posi-tive values of $[X]$. What does this imply about the shape of a plot of Y against $[X]$?

8.3   Derive an expression for the Hill coefficient in terms of K_1, K_2 and $[X]$ for a system that obeys equation 8.6. Show that in this case a definition of co-operativity in terms of h is identical to one in terms of K_1 and K_2. At what value of $[X]$ is h a maximum or minimum, and what is its extreme value?

8.4   Extension of the symmetry model with simple assumptions to allosteric inhibition and activation leads to a binding equation for X of the form of equation 8.10, with L replaced with a parameter L' that is directly proportional to $(1 + [M_1]/K_{M1})^4$, but inversely proportional to $(1 + [M_2]/K_{M2})^4$, where M_1 and M_2 are two different allosteric effectors and K_{M1} and K_{M2} are constants. Classify M_1 and M_2 as inhibitors or activators. For this model, does the homotropic co-operativity of X increase or decrease as the concentrations of (a) M_1, (b) M_2 increase?

Chapter 9

Fast reactions

9.1 Limitations of steady-state measurements

It is convenient to refer to the period in a reaction before the steady state is reached as the *transient phase*. This term is commonly used in physics and mathematics to describe terms of the form $A \exp(-t/\tau)$ that often occur in the solutions of differential equations. Such terms have finite and even very large values when t is small, but decay to zero as t is increased above τ, a constant called the *relaxation time* or *time constant*. Transients always occur in the kinetic equations that describe the progress of enzyme-catalysed reactions unless they are derived with the use of the steady-state assumption.

It is fairly obvious that experimental methods for investigating very fast reactions, with half-times of much less than 1 s, must be different from those used for slower reactions, because in most of the usual methods the time taken in mixing the reactants is of the order of seconds or greater. Less obviously, the kinetic equations required for fast reactions are also different, because in most enzyme-catalysed reactions the steady state is attained very rapidly (*see* Section 2.4) and can be considered to exist throughout the period of investigation, provided that this period does not include the first second after mixing. Consequently, most of the equations that have been discussed in this book have been derived with the use of the steady-state assumption. However, fast reactions are concerned, almost by definition, with the transient phase before the attainment of a steady state and cannot be described by steady-state rate equations. This chapter will be concerned with experimental and analytical aspects of the transient phase, but first it is useful to examine the reasons for making transient-phase measurements.

Although steady-state measurements have proved very useful for elucidating the mechanisms of enzyme-catalysed reactions, they suffer from the major disadvantage that, at best, the steady-state velocity of a multi-step reaction is the velocity of the slowest step, and steady-state measurements do not normally provide information about the

177

faster steps. Yet if the mechanism of an enzyme-catalysed reaction is to be understood, it is necessary to have information about steps other than the slowest. As discussed in Chapter 6, the experimenter has considerable freedom to alter the relative rates of the various steps in a reaction, by varying the concentrations of the substrates. Consequently it is often possible to examine more than one step of a reaction in spite of this limitation. However, isomerizations of intermediates along the reaction pathway cannot be separated in this way. To take a simple example:

$$E + A \underset{k_{-1}}{\overset{k_{+1}}{\rightleftharpoons}} EA \underset{k_{-2}}{\overset{k_{+2}}{\rightleftharpoons}} EP \underset{k_{-3}}{\overset{k_{+3}}{\rightleftharpoons}} E + P$$

The steady-state equation for this mechanism is equation 2.15, i.e.

$$v = \frac{(V^f a/K_m^A) - (V^r p/K_m^P)}{1 + (a/K_m^A) + (p/K_m^P)}$$

which contains only four parameters, and it is impossible to obtain from them the values of all of the six independent rate constants. This equation also applies to much more complex mechanisms in which the interconversion of EA and EP involves several intermediates. So steady-state measurements not only fail to provide any information about the individual steps; they also give no indication of how many steps there are. In general, as mentioned in Section 4.1, in any part of a reaction pathway that consists of a series of isomerizations of intermediates, all of the intermediates must be treated as a single species in steady-state kinetics. This is a severe limitation and provides the main justification for transient-state kinetics, which are subject to no such limitation.

The conveniently low rates observed in steady-state experiments are commonly achieved by working with very small concentrations of enzyme. This may be an advantage if the enzyme is very expensive or available in very small amounts, but it also means that all information about the enzyme is obtained at second hand, by observing effects on reactants, and not by observing the enzyme itself. If one wants to observe the enzyme itself, one must use it in reagent quantities so that it can be detected by spectroscopic or other techniques. This in practice means that the enzyme concentration is so high that steady-state methods cannot be used.

The advantages of transient-state methods may seem to make steady-state kinetics obsolete, but it is likely that steady-state investigations will continue to predominate for many years, for reasons that I shall now consider. First, the theory of the steady state is simpler, and steady-state measurements require less specialized equipment. In addition, the very small amounts of enzyme required in steady-state measurements allow them to be made on many enzymes for which transient-state experiments would be prohibitively expensive.

One should also be aware that the analysis of transient-state data suffers severely from a numerical difficulty known as *ill-conditioning*. This means that, *even in the absence of experimental error*, it is possible to fit experimental results with a wide range of constants and indeed of equations. This is illustrated in *Figure 9.1*, which shows

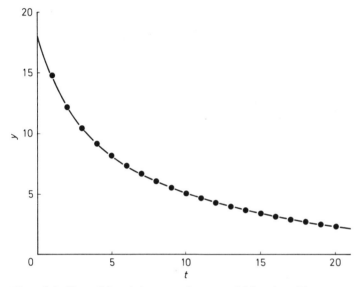

Figure 9.1   Ill-conditioned character of exponential functions. The points were calculated from $y = 5.1 \exp(-t/1.3) + 4.7 \exp(-t/4.4) + 9.3 \exp(-t/14.2)$, the line from $y = 7.32 \exp(-t/2.162) + 10.914 \exp(-t/12.86)$

a set of points and a line calculated from two different equations, both of the type commonly encountered in transient-state kinetics. The practical implication is that it is often impossible to extract all of the extra information that is theoretically present in transient-state measurements unless the various processes are very well separated on the time scale.

9.2 Active-site titration: 'burst' kinetics

In studies of the chymotrypsin-catalysed hydrolysis of nitrophenyl-ethyl carbonate, Hartley and Kilby (1954) observed that, although the release of nitrophenolate was almost linear, extrapolation of the line back to the product axis gave a positive intercept (*Figure 9.2*). Because the substrate was not a specific one for the enzyme and was consequently very poor, it was necessary to work with high enzyme concentrations and the intercept, which is known as a '*burst*' of product, was proportional to the enzyme concentration. This

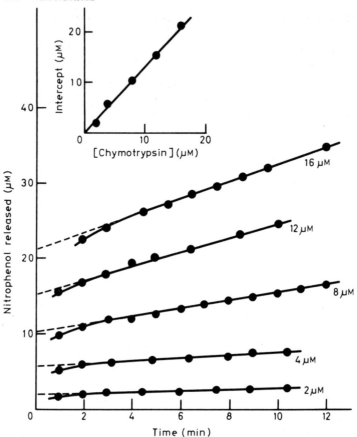

Figure 9.2 A 'burst' of product release. Data of Hartley and Kilby (1954) for the chymotrypsin-catalysed hydrolysis of nitrophenylethyl carbonate. The inset shows that the intercepts obtained by extrapolating the straight portions of the progress curves back to zero time are proportional to and almost equal to the enzyme concentration

suggested a mechanism in which the products were released in two steps, the nitrophenolate being released first:

$$E + S \underset{k_{-1}}{\overset{k_{+1}}{\rightleftharpoons}} ES \overset{k_{+2}}{\longrightarrow} EQ \overset{k_{+3}}{\longrightarrow} E + Q$$
$$\searrow$$
$$P$$

If the final step is rate-limiting, that is, if k_{+3} is small compared with $k_{+1}s$, k_{-1} and k_{+2}, then the enzyme will exist almost entirely as EQ in the steady state. It is not necessary for EQ to be formed before P can be released, however, and so in the transient phase P can be released at a rate much greater than the steady-state rate. It might be expected that the amount of P released in the burst would be equal, and not merely proportional, to the amount of enzyme. This is accurately true only if k_{+3} is very much smaller than the other rate constants; otherwise, the burst is smaller than the stoicheiometric amount, as will now be shown, following a derivation based on that of Gutfreund (1955).

If s is large enough to be treated as a constant during the time period considered, and if $k_{+1}s$ is large compared with $(k_{-1} + k_{+2} + k_{+3})$, then very shortly after mixing the system effectively simplifies to

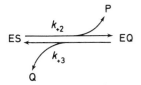

because the reaction $E + S \rightarrow ES$ can be regarded as instantaneous and irreversible, and the concentration of free enzyme becomes negligible. This is then a simple reversible first-order reaction (cf. Section 1.4), with the solution

$$[ES] = \frac{e_0\{k_{+3} + k_{+2} \exp [-(k_{+2} + k_{+3})t]\}}{k_{+2} + k_{+3}}$$

$$[EQ] = \frac{k_{+2}e_0\{1 - \exp [-(k_{+2} + k_{+3})t]\}}{k_{+2} + k_{+3}}$$

Expressions for the rates of release of the two products are readily obtained by multiplying these two equations by k_{+2} and k_{+3} respectively:

$$\frac{dp}{dt} = k_{+2}[ES] = \frac{k_{+2}e_0\{k_{+3} + k_{+2} \exp [-(k_{+2} + k_{+3})t]\}}{k_{+2} + k_{+3}}$$

$$\frac{dq}{dt} = k_{+3}[EQ] = \frac{k_{+2}k_{+3}e_0\{1 - \exp [-(k_{+2} + k_{+3})t]\}}{k_{+2} + k_{+3}}$$

In the steady state, i.e. when t is large, the exponential term becomes negligible and the two rates become equivalent:

$$\frac{dp}{dt} = \frac{dq}{dt} = \frac{k_{+2}k_{+3}e_0}{k_{+2} + k_{+3}}$$

In the transient phase, however, dp/dt is initially much larger than dq/dt, so that whereas P displays a 'burst', Q displays a 'lag' when the linear parts of the progress curves are extrapolated back to zero time. The magnitude of the burst can be calculated by integrating the expression for dp/dt and introducing the condition $p = 0$ when $t = 0$:

$$p = \frac{k_{+2}k_{+3}e_0 t}{k_{+2} + k_{+3}} + \frac{k_{+2}^2 e_0 \{1 - \exp[-(k_{+2} + k_{+3})t]\}}{(k_{+2} + k_{+3})^2}$$

An expression for the steady-state portion of the progress curve can now be obtained by omitting the exponential term:

$$p = \frac{k_{+2}k_{+3}e_0 t}{k_{+2} + k_{+3}} + \frac{k_{+2}^2 e_0}{(k_{+2} + k_{+3})^2}$$

This is the equation for a straight line and the intercept on the p axis gives π, the magnitude of the burst:

$$\pi = \frac{k_{+2}^2 e_0}{(k_{+2} + k_{+3})^2} = e_0/(1 + k_{+3}/k_{+2})^2 \tag{9.1}$$

Thus the burst in P is *not* equal to the enzyme concentration, but approximates to it if $k_{+2} \gg k_{+3}$. This equation implies that the burst can never exceed the enzyme concentration, but extrapolation of the 'linear' portion of a progress curve can sometimes yield an over-estimate of the true burst size if the velocity is not accurately constant in the steady state but decays at an appreciable rate. One can avoid this type of error by ensuring that the progress curve is truly straight during the steady-state phase.

The discovery of burst kinetics has led to an important method for titrating enzymes. It is generally difficult to obtain an accurate measure of the molarity of an enzyme: rate assays provide concentrations in activity units or katals per millilitre, which are adequate for comparative purposes but do not provide true concentrations unless they have been calibrated in some way; most other assays are really protein assays and are therefore very unspecific unless the enzyme is known to be pure and fully active. However, equation 9.1 shows that, if a substrate can be found for which k_{+3} is either very small or zero, then the burst π is both well-defined and equal to the concentration of active sites. The substrates of chymotrypsin that were examined originally, p-nitrophenylethyl carbonate and p-nitrophenyl acetate, showed values of k_{+3} that were inconveniently large, but subsequently Schonbaum, Zerner and Bender (1961) found that under suitable conditions *trans*-cinnamoylimidazole gave excellent results. This compound reacts rapidly with chymotrypsin at pH 5.5 to give imidazole and *trans*-cinnamoylchymotrypsin, but no further reaction occurs readily, i.e. $k_{+3} \simeq 0$. So measurement of the amount of imidazole released by a solution of chymotrypsin provides a measure of the amount of enzyme. Suitable titrants for a number of other enzymes have also been found (*see*, for example, Bender *et al.*, 1966).

Active-site titration by means of burst measurements differs from rate assays in being relatively insensitive to changes in the rate constants: a rate assay demands precisely defined conditions of pH, temperature, buffer composition, etc. if it is to be reproducible, but the magnitude of a burst is unaffected by relatively large changes in k_{+2}, such as might result from chemical modification of the enzyme, unless k_{+2} is decreased until it is comparable in magnitude to k_{+3}. Thus chemical modification alters the molarity of an enzyme, as measured by this technique, either to zero or not at all. For this reason, enzyme titration has also been called an '*all-or-none*' assay (Koshland, Strumeyer and Ray, 1962).

9.3 Flow methods

Steady-state experiments are usually carried out over a time scale of several minutes at least. They have not required the development of special equipment, because, in principle, any method that permits the analysis of a reaction mixture at equilibrium can be adapted to allow analysis during the course of reaction. In the study of fast reactions, however, the short time periods involved have required specially designed apparatus that cannot be regarded simply as an obvious extension of that used in steady-state experiments.

For processes with time constants of the order of milliseconds or greater, the techniques that have been devised are mainly *flow methods*, derived ultimately from the *continuous-flow* method used by Hartridge and Roughton (1923) for measuring the rate of combination of oxygen with haemoglobin. In this method the reaction was initiated by forcibly mixing the two reagents, reduced haemoglobin and oxygenated buffer, so that the mixture was caused to move rapidly and turbulently down a long (1 m) tube. For a constant flow rate, the mixture observed at any point along the tube had a constant age that depended on the flow rate and the distance from the mixing chamber. So by making measurements at several points along the tube one could obtain a progress curve for the early stages of the reaction.

The date of the experiments of Hartridge and Roughton (1923) is of particular interest because it precedes the commercial availability of automatic devices for measuring light intensity. It illustrates how scientific ingenuity can overcome seemingly impossible obstacles, in this case how to measure changes that occur in a time scale of milliseconds with equipment that requires several seconds of manual adjustment for each measurement. Nevertheless, the continuous-flow method had severe drawbacks that in practice restricted its use to the study of haemoglobin.

Various major improvements by Millikan (1936a, b), Chance (1940, 1951), Gibson and Milnes (1964) and others led to the development of the *stopped-flow method*, which has become the most widely used method for the study of fast reactions. The essentials of the apparatus are shown in *Figure 9.3*, and consist of: (a) two drive

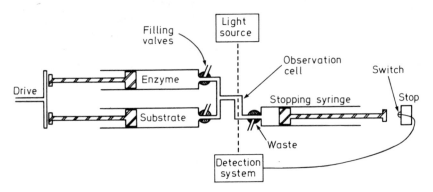

Figure 9.3 Essentials of the stopped-flow apparatus

syringes containing the reacting species, (b) a mixing device, (c) an observation cell, (d) a stopping syringe, and (e) a detecting and recording system that is capable of responding very rapidly. The reaction is started by pushing the plungers of the two drive syringes simultaneously. This causes the two reactants to mix and the mixture is forced through the observation cell and into the stopping syringe. A short movement of the plunger of the stopping syringe brings it to a mechanical stop, which prevents further mixing and simultaneously activates the detection and recording system. The *dead time* of this apparatus, that is, the time that inevitably elapses between the first mixing of reactants and the arrival of the mixture in the observation cell, is of the order of 1 ms.

In its usual form, the stopped-flow method requires a spectroscopic device for following the course of the reaction. This makes it particularly useful for the study of reactions that are accompanied by a large change in absorbance at a convenient wavelength, such as a dehydrogenase-catalysed reaction in which NAD^+ is reduced to NADH. The method is not, however, restricted to such cases because other detection methods are available. For example, many enzyme-catalysed reactions are accompanied by the release or uptake of protons, which can be detected optically by including a pH indicator in the reaction mixture, an approach that originated at the time of the continuous-flow method (Brinkman, Margaria and Roughton, 1933).

In other cases, changes in fluorescence during the reaction may be exploited (e.g. Hastings and Gibson, 1963).

There are sometimes doubts about the chemical nature of the events observed spectroscopically in the stopped-flow method (Porter, 1967). These can in principle be overcome by use of the *quenched-flow method*. In this method the reaction is stopped ('quenched') shortly after mixing, either by a second mixing in which enzyme activity is rapidly destroyed by a denaturing agent such as trichloroacetic acid, or by very rapid cooling to a temperature at which the reaction rate is negligible. By varying the time between the initial mixing and the subsequent quenching, one can obtain a series of samples that can be analysed by chemical or other means, from which a record of the chemical progress of the reaction can be reconstructed.

The quenched-flow method requires much larger amounts of enzyme and other reagents than the stopped-flow method, because each run yields only a single point on the time course, whereas each stopped-flow run yields a complete time course. In consequence, it is often appropriate to apply the quenched-flow method only after preliminary stopped-flow experiments have established the proper questions to be asked. Consider, for example, the data shown in *Figure 9.4*, which were obtained by Eady, Lowe and Thorneley (1978) in studies of the $MgATP^{2-}$ dependence of nitrogenase. This enzyme, which is responsible for the biological fixation of molecular nitrogen, consists of two redox proteins, the Fe protein and the Mo–Fe protein. $MgATP^{2-}$ is required for the transfer of electrons from the Fe protein to the Mo–Fe protein and is hydrolysed to $MgADP^{-}$ and inorganic phosphate during the reaction. The oxidation of the Fe protein can be observed directly in the stopped-flow spectrophotometer at 420 nm; this reaction is initiated by mixing with $MgATP^{2-}$ and shows a single relaxation with $\tau = 42 \pm 3$ ms (*Figure 9.4a*). This observation does not by itself establish that hydrolysis of $MgATP^{2-}$ and electron transfer are directly coupled; instead, $MgATP^{2-}$ might merely be an activator of the Fe protein. To resolve this question the rate of hydrolysis had to be directly measured, which was done by measuring the production of inorganic phosphate by the quenched-flow method. This process proved to have $\tau = 44 \pm 4$ ms (*Figure 9.4b*), indistinguishable from the value measured in the stopped-flow method. Thus the two reactions do appear to be synchronous.

In the earliest versions of the quenched-flow method, the time between mixing and quenching was varied by varying the physical design of the apparatus, that is, by varying the flow rate and the length of tube between the two mixing devices. In practice there were severe restrictions on the time scales that could be used and

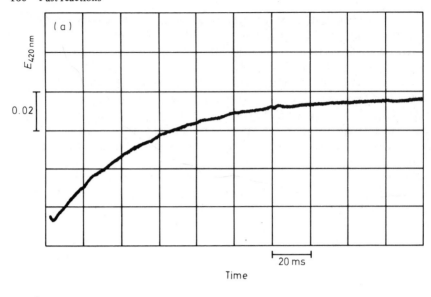

Time

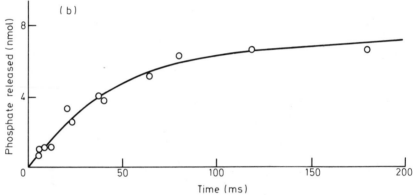

Time (ms)

Figure 9.4 (a) Stopped-flow, and (b) quenched-flow data for the reaction catalysed by nitrogenase from *Klebsiella pneumoniae* (Eady, Lowe and Thorneley, 1978). The stopped-flow trace records the electron transfer between the two component proteins of nitrogenase, whereas the quenched-flow observations measure the rate of appearance of inorganic phosphate, i.e. the hydrolysis of ATP. The equality of the time constants shows that the two processes are coupled

the method was impractical for general use. Many of the problems have been overcome in an ingenious modification known as *pulsed quenched flow*, which has been described by Fersht and Jakes (1975). In this arrangement, the reaction is initiated exactly as in a stopped-flow experiment, and a second set of syringes is used for quenching. These are actuated automatically after a preset time has elapsed from the original mixing. The time between mixing and quenching is controlled electronically and does not depend on the physical dimensions of the apparatus. Because of the absence of any need for long tubes, this system is also much more economical of reagents than the conventional quenched-flow method.

An alternative to quenching as a means of obtaining more information about the chemical nature of the processes seen in the stopped-flow apparatus has been described by Hollaway and White (1975). By using a rapidly scanning spectrophotometer capable of recording spectra over a range of about 200 nm at the rate of 800 spectra per second, one can acquire detailed spectroscopic information about the progress of a reaction. *Figure 9.5* shows the appearance of the

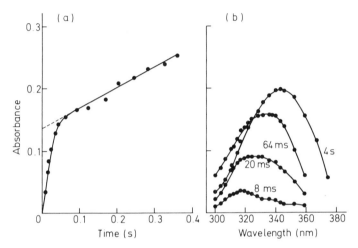

Figure 9.5 Appearance of the NADH spectrum in the reaction catalysed by horse-liver alcohol dehydrogenase (Hollaway and White, 1975). The plot in (a) shows observations at a single wavelength, 340 nm, and is similar to what one would observe in a conventional stopped-flow experiment. The form of the curve suggests a rapid release of NADH in the transient phase, followed by slower steady-state release. However, the individual spectra recorded at various times after mixing, and shown in (b), show that the spectrum of the NADH produced in the transient phase is different from that of free NADH, with $\lambda_{max} = 320$ nm instead of 340 nm. The authors interpreted this result as evidence that the NADH produced in the transient phase was enzyme-bound and that no rapid release of NADH could take place before the rate-limiting step of the catalysis

NADH spectrum in the transient phase of the reaction catalysed by alcohol dehydrogenase from horse liver. Although the trace at 340 nm, similar to what one would see in a conventional stopped-flow experiment, does indicate a rapid burst of NADH followed by the slower steady-state reaction (cf. *Figure 9.2*), it fails to show that the spectrum of the NADH formed in the burst is not the same as that of NADH free in solution; the maximum absorbance occurs at around 320 nm instead of 340 nm. Hollaway and White interpreted this spectrum as evidence that the NADH formed in the burst is bound to the enzyme, and thus excluded any mechanism that would allow rapid NADH release before the rate-limiting step of the reaction.

This method has not yet found widespread application, probably because of the elaborate apparatus required, but there can be no doubt of its potential usefulness as a method that provides much more chemical information than conventional stopped flow, and is much quicker, more convenient to use, and more economical of materials, than quenched flow.

9.4 Relaxation methods

Mixing of reagents cannot be done effectively in less than about 0.2 ms, and stopping the flow of a mixture through an apparatus requires about 0.5 ms. (Although one can conceive of a more rapid stopping, in practice shock waves would be created that would generate artefactual transients in the detection system.) Quenching, either chemically or by cooling, also requires finite time. There is therefore a lower limit of about 0.5 ms to the dead time that it is possible to achieve with flow methods, and it is unlikely that improved design will decrease this appreciably. So processes that are virtually complete within 0.5 ms cannot be observed by flow methods. This is a severe restriction for the enzymologist, because most enzyme-catalysed reactions contain some such processes, and for several the complete catalytic cycle requires less than 1 ms. For example, in a recent compilation of results, Fersht (1977) lists five enzymes for which the value of k_{cat} is of the order of 10^3 s^{-1} or greater.

These considerations have led to the development of *relaxation methods* (Eigen, 1954) for studying very fast processes; these methods do not require any mixing of reagents. Instead, a mixture at equilibrium is subjected to a *perturbation* that alters the equilibrium constant, and one then observes the system proceeding to the new equilibrium, a process known as *relaxation*. Various types of perturbation are possible, but the most commonly used is to pass a large electric discharge through the reaction mixture, which brings about an increase in temperature of the order of 10 °C in a period of the order of 1 μs. This is known as the *temperature-jump* method. Other perturbations, such as a pressure jump, are less useful because they produce much smaller changes in the kinetic parameters for a comparable input of energy.

The perturbation caused by a temperature jump is not instantaneous, but takes a finite time of the order of 1 μs. It can be regarded as instantaneous only if one confines attention to processes that occur more than 1 μs after the start of the perturbation. It is unlikely that appreciably shorter times can be achieved with irreversible perturbations, but very much faster processes can be studied by means of sinusoidal perturbations. For example, ultrasonic waves, with frequency as high as 10^{11} s^{-1}, are accompanied by local fluctuations in temperature and pressure as they are propagated through a medium. These fluctuations produce oscillations in the values of all the rate constants of the system, and study of the absorption of ultrasonic energy by a reaction mixture yields information about these rate constants (Hammes and Schimmel, 1970). Enzyme systems are generally too complicated for direct application of this method. Nonetheless, study of simple systems has provided valuable information for the enzymologist: for example, the work of Burke, Hammes and Lewis (1965) on poly-L-glutamate has shown that a major conformational change of a macromolecule, the helix–coil transition, can occur at a rate of 10^5–2×10^7 s^{-1}. Obviously, conformational changes in enzymes do not have to occur at the same rate, but similar rates must be possible, and thus the need for a fast conformational change does not provide any objection to a proposed mechanism for enzymic catalysis.

One disadvantage of observing relaxation of a system to equilibrium is that the transient species of particular importance in the catalytic process may occur only at very low concentrations when the system is close to equilibrium. This difficulty may be overcome by combining the stopped-flow and temperature-jump methods, and stopped-flow apparatus that includes a temperature-jump facility is now commercially available. In this application the reactants are mixed as in a stopped-flow experiment, and subsequently subjected to a temperature jump after a steady state has been attained. Relaxation to the new steady state characteristic of the higher temperature is then observed. This sort of experiment permits the observation of processes in the early phase of reaction that are too fast for the conventional stopped-flow method, because the reactants are already mixed when the temperature jump occurs.

9.5 Transient-state kinetics of systems far from equilibrium

In Section 2.4, I discussed the validity of the steady-state assumption by deriving an equation for the kinetics of the two-step Michaelis–Menten mechanism without assuming a steady state. This derivation was only made possible, however, by treating the substrate concentration as a constant, which was clearly not exactly correct. In fact, if no assumptions are made, solution of the differential equations is impossible for nearly all mechanisms of enzymic catalysis; approximations must always be made, therefore, if any analysis is to be possible. In transient-state experiments one usually tries to set up conditions such that the mechanism approximates to a sequence of first-order steps, because this is the most general sort of mechanism for which an exact solution exists.

The following two-step mechanism will serve to illustrate the way in which sequences of first-order steps can be analysed:

$$X_0 \xrightleftharpoons[k_{-1}]{k_{+1}} X_1 \xrightleftharpoons[k_{-2}]{k_{+2}} X_2 \qquad (9.2)$$

The system is defined by a *conservation equation* and three rate equations. The conservation equation,

$$x_0 + x_1 + x_2 = x_{tot}$$

ensures that the requirements of stoicheiometry are met, by requiring the sum of the three concentrations to be a constant, x_{tot}. The three rate equations are as follows:

$$dx_0/dt = -k_{+1}x_0 + k_{-1}x_1 \qquad (9.3)$$

$$dx_1/dt = k_{+1}x_0 - (k_{-1} + k_{+2})x_1 + k_{-2}x_2 \qquad (9.4)$$

$$dx_2/dt = k_{+2}x_1 - k_{-2}x_2$$

Any one of these three equations is redundant, as their sum is simply the first derivative of the conservation equation:

$$\frac{dx_0}{dt} + \frac{dx_1}{dt} + \frac{dx_2}{dt} = 0$$

Solution of a set of three differential equations in three unknown concentrations is most easily achieved by eliminating two of the concentrations to produce a single differential equation in one unknown. First we may eliminate x_2 by expressing its concentration in terms of those of the other two species,

$$x_2 = x_{tot} - x_0 - x_1$$

and substituting in equation 9.4:

$$dx_1/dt = k_{+1}x_0 - (k_{-1} + k_{+2})x_1 + k_{-2}(x_{tot} - x_0 - x_1)$$

$$= (k_{+1} - k_{-2})x_0 - (k_{-1} + k_{+2} + k_{-2})x_1 + k_{-2}x_{tot}$$

Differentiating equation 9.3, we have

$$\frac{d^2x_0}{dt^2} = -k_{+1}\frac{dx_0}{dt} + k_{-1}\frac{dx_1}{dt}$$

$$= -k_{+1}\frac{dx_0}{dt} + k_{-1}(k_{+1} - k_{-2})x_0 - k_{-1}(k_{-1} + k_{+2} + k_{-2})x_1$$
$$+ k_{-1}k_{-2}x_{tot} \qquad (9.5)$$

Next x_1 may be eliminated by rearranging equation 9.3:

$$k_{-1} x_1 = \frac{dx_0}{dt} - k_{+1} x_0$$

and substituting in equation 9.5:

$$\frac{d^2 x_0}{dt^2} = -k_{+1} \frac{dx_0}{dt} + k_{-1}(k_{+1} - k_{-2})x_0$$

$$- (k_{-1} + k_{+2} + k_{-2})\left(\frac{dx_0}{dt} - k_{+1} x_0\right) + k_{-1} k_{-2} x_{tot}$$

which, after rearranging, becomes

$$\frac{d^2 x_0}{dt^2} + (k_{+1} + k_{-1} + k_{+2} + k_{-2}) \frac{dx_0}{dt}$$

$$+ (k_{-1} k_{-2} + k_{+1} k_{+2} + k_{+1} k_{-2})x_0 = k_{-1} k_{-2} x_{tot}$$

This is now a linear differential equation in x_0 and has the standard form

$$\frac{d^2 x_0}{dt^2} + P \frac{dx_0}{dt} + Q x_0 = R$$

and has the solution

$$x_0 = x_{0\infty} + A_{01} \exp(-t/\tau_1) + A_{02} \exp(-t/\tau_2)$$

in which $x_{0\infty}$ is the value of x_0 at equilibrium, A_{01} and A_{02} are constants of integration defined by the initial state of the system known as the *amplitudes* of the two exponential terms, and τ_1 and τ_2 are the corresponding *relaxation times* or *time constants*. In common with most elementary accounts of relaxation kinetics, this discussion will concentrate on the information content of the relaxation times, which are rather simpler to treat mathematically than the amplitudes. Nonetheless, it should not be forgotten that amplitudes provide a potentially rich source of additional information. Because relaxation methods involve input of energy, usually in the form of heat, the relaxation amplitude depends on the thermodynamic properties of the system, such as the enthalpy of reaction ΔH. Consequently, amplitude measurements can provide more accurate information about these thermodynamic properties than is available from ordinary measurements. Thusius (1973) describes how this is done for the simple case of 1 : 1 complex formation, and provides references to other sources of information.

In the above expression, the values of τ_1 and τ_2 are given by

$$1/\tau_1 = \tfrac{1}{2}[P + (P^2 - 4Q)^{\frac{1}{2}}]$$
$$1/\tau_2 = \tfrac{1}{2}[P - (P^2 - 4Q)^{\frac{1}{2}}]$$

The solutions for x_1 and x_2 are of the same form, with the same pair of relaxation times, but with different amplitudes.

If $k_{-1}k_{+2}$ is small compared with $(k_{-1}k_{-2} + k_{+1}k_{+2} + k_{+1}k_{-2})$, the expressions for $1/\tau_1$ and $1/\tau_2$ simplify to $(k_{+1} + k_{-1})$ and $(k_{+2} + k_{-2})$, not necessarily respectively: $1/\tau_1$ is always the larger, $1/\tau_2$ always the smaller. Even if this simplification is not permissible, the expressions for the sum and product of $1/\tau_1$ and $1/\tau_2$ are fairly simple:

$$\frac{1}{\tau_1} + \frac{1}{\tau_2} = k_{+1} + k_{-1} + k_{+2} + k_{-2} = P \tag{9.6}$$

$$1/\tau_1\tau_2 = k_{+1}k_{-2} + k_{+1}k_{+2} + k_{+1}k_{-2} = Q \tag{9.7}$$

This example illustrates several points that apply more generally. Any mechanism that consists of a sequence of n steps that are first order in both directions can be solved exactly. The solution allows the concentration of any reactant or intermediate to be expressed as the sum of its value at equilibrium and n exponential terms, with amplitudes that are different for each reactant and relaxation times that are the same for each reactant. In favourable cases the relaxation times for some or all of the exponentials can be associated with particular steps in the mechanism; when this is true the reciprocal of the relaxation time is equal to the sum of the rate constants for the forward and reverse reactions. One other general point, which is not obvious from the above analysis, is that reactants separated from the rest of the mechanism by irreversible steps display simpler relaxation spectra than other reactants, because some of the amplitudes are zero. Consider for example the following five-step mechanism, in which two of the steps are irreversible:

$$X_0 \rightleftarrows X_1 \longrightarrow X_2 \rightleftarrows X_3 \longrightarrow X_4 \rightleftarrows X_5$$

In principle, each reactant should display five relaxation times, and this is indeed what one would expect to observe for X_4 and X_5. But X_2 and X_3 are isolated from the last step by the irreversible fourth step, and they therefore have one zero amplitude, and only four relaxation times. X_0 and X_1 are isolated from the rest of the mechanism by the irreversible second step, and consequently they each have three zero amplitudes, and only two relaxation times. In addition, regardless of the existence of irreversible steps, the total number of relaxations observed cannot exceed the predicted number, but it is often less, either because processes with similar relaxation times are not resolved, or because some of the amplitudes are too small to be detected.

All mechanisms for enzymic catalysis include at least one second-order step. However, any such step can be made to follow first-order kinetics with respect to time (i.e. pseudo-first-order kinetics: cf. Section 1.1), by ensuring that one of the two reactants involved is in large excess over the other. It follows that at least one of the observed relaxation times contains a pseudo-first-order rate constant and thus the expression for it includes a concentration dependence. This is very useful as it allows measured relaxation times to be assigned to particular steps. Consider, for example, the following mechanism, which represents half of a substituted-enzyme reaction (Section 6.2) studied in the absence of the second substrate:

$$E \underset{k_{-1}}{\overset{k_{+1}a}{\rightleftharpoons}} EA \xrightarrow{k_{+2}} E' + P$$

Equations 9.6 and 9.7 take the form

$$\frac{1}{\tau_1} + \frac{1}{\tau_2} = k_{+1}a + k_{-1} + k_{+2}$$

$$1/\tau_1\tau_2 = k_{+1}k_{+2}a$$

So a plot of the sum of the two reciprocal relaxation times against a yields a straight line of slope k_{+1} and intercept $(k_{-1} + k_{+2})$ on the ordinate, and a plot of their product against a yields a straight line through the origin with slope $k_{+1}k_{+2}$. Thus one can calculate all three rate constants from measured values of the relaxation times.

9.6 Simplification of complex mechanisms

Although a system of n unimolecular steps can in principle be analysed exactly, regardless of the value of n, in practice it is diffi-cult to resolve exponential processes unless they are well separated on the time axis. Consequently the number of transients detected may well be less than the number present. The degree of separation necessary for resolution depends on the amplitudes, but some gener-alizations are possible. If two processes have amplitudes of opposite sign they are relatively easy to resolve, even if the relaxation times are within a factor of 2. The reason for this is fairly obvious: if a transient appears and then disappears there can be no doubt that at least two processes are involved. For transients with amplitudes of the same sign, resolution is considerably more difficult, because the decay curve is monotonic and unless the faster process has a consider-ably larger amplitude than the slower one its presence may pass unnoticed (cf. *Figure 9.1*, p. 179).

In general, slow relaxations are easier to measure than fast ones, because one can follow them in a time scale in which all of the faster ones have decayed to zero. In principle, therefore, one can examine the faster processes by subtracting out the slower ones. Consider, for example, the equation

$$x = x_\infty + A_1 \exp{(-t/\tau_1)} + A_2 \exp{(-t/\tau_2)}$$

in which τ_2 is larger than τ_1 by a factor of at least 10. One can evaluate x_∞ by allowing the reaction to proceed to equilibrium, and then determine A_2 and τ_2 by making measurements over a period from about $0.5\tau_2$ to $5\tau_2$. Then, subtracting the calculated value of $x_\infty + A_2 \exp{(-t/\tau_2)}$ from x for each value of t in the early part of the progress curve, one should obtain data for a single relaxation $A_1 \exp{(-t/\tau_1)}$. In this process, which is known as *peeling*, the errors accumulate as one proceeds: any inaccuracy in x_∞ contributes to the errors in A_2 and τ_2, and any inaccuracies in x_∞, A_2 and τ_2 contribute to the errors in A_1 and τ_1. So although in principle one can separate any number of relaxations by this method, in practice the faster processes are much less well-defined than the slower ones. For practical purposes, therefore, it is advisable to create experimental conditions in which the number of relaxations is as small as possible. An example of this was seen in Section 9.1: although the three-step mechanism used to explain the burst of product release should in principle give rise to two relaxations, the number was decreased to one by the use of such a high substrate concentration that the first relaxation could be treated as instantaneous.

Considered from the point of view of the enzyme, enzyme-catalysed reactions are usually cyclic; that is, the first reactant, the free enzyme, is also the final product. This does not prevent solution of the differential equations (provided, as before, that every step is first-order or pseudo-first-order), but it does lead to more complicated transient kinetics than one obtains for non-cyclic reactions. This is because the system relaxes to a steady state rather than to equilibrium. It is therefore useful to simplify matters by removing the cyclic character from the reaction. There are various ways of doing this, of which conceptually the simplest is to choose a substrate for which the steady-state rate is so small that it can be ignored. In effect, this is what one does in using an active-site titrant (Section 9.2). It does, however, have the disadvantage that one must usually study the enzyme with an unnatural substrate.

A different approach is to carry out a *single-turnover experiment*, in which the rate is limited by substrate, not enzyme, i.e. $e_0 \gg s_0$. In this case the Michaelis–Menten mechanism becomes

$$S \underset{k_{-1}}{\overset{k_{+1}e}{\rightleftharpoons}} ES \underset{k_{-2}e}{\overset{k_{+2}}{\rightleftharpoons}} P$$

This is of the form of equation 9.2 because the reaction must stop when the substrate is consumed and so no recycling of enzyme can take place.

In reactions with more than one substrate, apart from hydrolytic reactions, one can prevent recycling of enzyme by omitting one substrate from the reaction mixture. This is particularly useful for enzymes that follow a substituted-enzyme mechanism, because some chemical reaction occurs, and is potentially measurable, even in incomplete mixtures. By studying the partial reactions catalysed by glutamate–aspartate transaminase, mainly by the temperature-jump method, Hammes and Fasella (1962) were able to assign values to 10 of the 12 rate constants that occur in the mechanism (*Figure 9.6*).

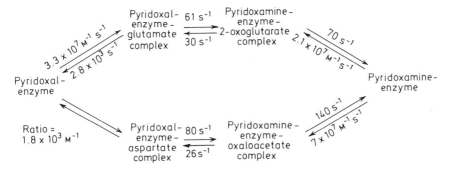

Figure 9.6   The mechanism of the reaction catalysed by glutamate–aspartate transaminase, with the rate constants assigned by Hammes and Fasella (1962)

Transaminases form a particularly attractive class of enzymes for such studies on account of the easily monitored spectral changes in the coenzyme, pyridoxal phosphate, during the reaction.

The treatment of enzymes that follow a ternary-complex mechanism is less straightforward, because a complete reaction mixture is normally required for any chemical change to occur. Nonetheless Pettersson (1976) has provided a rigorous treatment of the transient kinetics of ternary-complex reactions, and applied it to resolve some ambiguities in the reaction catalysed by alcohol dehydrogenase from horse liver (Kvassman and Pettersson, 1976). Single-turnover experiments can be carried out by keeping one substrate (not both) at a much lower concentration than the enzyme.

One of the attractive features of fast-reaction kinetics is that it can often provide conceptually simple information about mechanisms without any of the algebraic complexity that can hardly be avoided in steady-state work. An obvious example of this was the original 'burst' experiment of Hartley and Kilby (1954) (*see* Section 9.2): in this experiment the order of product release was established with a high degree of certainty by the observation that one product was released in a burst. Strictly, one ought to show that the other product does not also show a burst, because in principle there might be a burst in both products if the last step of the reaction were a rate-limiting isomerization of the free enzyme back to its original active form. Similarly, one can deduce the order of addition of substrates in a ternary-complex mechanism by varying the combinations of reagents in the syringes in a stopped-flow experiment. If there is a compulsory order of addition, the trace seen when the enzyme is premixed with the substrate that binds first is likely to be simpler than that seen when the enzyme is premixed with the substrate that binds second. In the former case the first complex is already formed when the syringes are actuated, but in the latter case pre-mixing with the second substrate achieves nothing because no reaction can take place until the enzyme is exposed to the first substrate.

9.7 Kinetics of systems close to equilibrium

In the temperature-jump apparatus the perturbation of the equilibrium constant is not usually large enough to establish a state in which the system is far from equilibrium. As a result, analysis of the kinetics is fairly simple and there is no requirement for all higher-order steps to be made pseudo-first-order. The reason for this is that the terms in products of concentrations that render the differential equations insoluble can be neglected in systems that are close to equilibrium. A simple binding reaction illustrates this:

$$
E \quad + \quad S \underset{k_{-1}}{\overset{k_{+1}}{\rightleftharpoons}} ES
$$

$$
e_\infty + \Delta e \qquad s_\infty + \Delta s \qquad\qquad x_\infty + \Delta x
$$

If e_∞, s_∞ and x_∞ are the final equilibrium concentrations of E, S and ES respectively, and k_{+1} and k_{-1} are the rate constants at the higher temperature, that is, those that define the system *after* perturbation, then the instantaneous concentrations can be represented as $(e_\infty + \Delta e)$, $(s_\infty + \Delta s)$ and $(x_\infty + \Delta x)$ respectively. Hence the rate is given by

$$
dx/dt = k_{+1}(e_\infty + \Delta e)(s_\infty + \Delta s) - k_{-1}(x_\infty + \Delta x)
$$

But $d\,\Delta x/dt = dx/dt$ and, by the stoicheiometry of the reaction, $\Delta e = \Delta s = -\Delta x$, and so

$$
d\,\Delta x/dt = k_{+1}(e_\infty - \Delta x)(s_\infty - \Delta x) - k_{-1}(x_\infty + \Delta x)
$$

$$
= k_{+1}e_\infty s_\infty - k_{-1}x_\infty - [k_{+1}(e_\infty + s_\infty) + k_{-1}]\,\Delta x + (\Delta x)^2
$$

$$
(9.8)
$$

As it stands, this is a non-linear differential equation and cannot be solved. But the term that makes it non-linear is $(\Delta x)^2$ and, provided the system is close to equilibrium, as assumed initially, this term can be neglected. In addition, the net rate at equilibrium is zero, by definition, and so $k_{+1} e_\infty s_\infty - k_{-1} x_\infty = 0$. So equation 9.8 simplifies to

$$d \Delta x/dt = -[k_{+1} (e_\infty + s_\infty) + k_{-1}] \Delta x$$

This is a simple linear differential equation that can be directly solved by separating the variables and integrating (cf. equation 1.1 in Section 1.1), to give

$$\Delta x = \Delta x_0 \exp \{ -[k_{+1} (e_\infty + s_\infty) + k_{-1}] t\}$$

where Δx_0 is the magnitude of the perturbation when $t = 0$. Thus, provided that the initial perturbation is small, the relaxation of a single-step reaction is described by a single exponential term with a relaxation time τ given by

$$1/\tau = k_{+1} (e_\infty + s_\infty) + k_{-1}$$

As e_∞, s_∞ and k_{-1}/k_{+1} can normally be measured independently, measurement of τ permits individual values to be assigned to k_{+1} and k_{-1}.

A similar analysis can be applied to any mechanism close to equilibrium, without regard to whether the individual steps are first-order or not, because the non-linear terms of the type $(\Delta x)^2$ can always be neglected. In general, a mechanism with n steps provides a solution with n exponential terms, though this number may be decreased by thermodynamic constraints on the allowed values for the rate constants. In addition, the number of relaxations observed experimentally is often less than the theoretical number, on account of failure to detect relaxations that have small amplitudes or are poorly resolved from others.

Although the system relaxing to a steady state considered at the end of Section 9.4 is not strictly a system close to equilibrium, it can be analysed in a similar way. Consider, for example, the Michaelis–Menten mechanism:

$$\mathrm{E} \quad + \quad \mathrm{S} \underset{k_{-1}}{\overset{k_{+1}}{\rightleftharpoons}} \mathrm{ES} \xrightarrow{k_{+2}} \mathrm{E} + \mathrm{P}$$

$$e_{ss} + \Delta e \qquad\qquad s_0 \qquad\qquad x_{ss} + \Delta x$$

subjected to a small perturbation from the steady state. The relaxation rate is given by

$$d \Delta x/dt = k_{+1} (e_{ss} + \Delta e) s_0 - (k_{-1} + k_{+2})(x_{ss} + \Delta x)$$

After introducing the steady-state condition $k_{+1} e_{ss} s_0 = (k_{-1} + k_{+2}) x_{ss}$, and the stoicheiometry requirement $\Delta e = -\Delta x$, this simplifies to

$$d \, \Delta x / dt = -(k_{+1} s_0 + k_{-1} + k_{+2}) \Delta x$$

which is readily integrable to give a solution consisting of a single exponential term:

$$\Delta x = \Delta x_0 \exp \left[-(k_{+1} s_0 + k_{-1} + k_{+2}) t \right]$$

For a more advanced and more detailed account of relaxation kinetics of enzymes, see Hammes and Schimmel (1970).

Problems

9.1 An enzyme of molecular weight 50 000 is studied in a single-turnover experiment at a substrate concentration of 1 μM. Measurement of the rate of product appearance at an enzyme concentration of 1 mg/ml reveals two exponential processes, with relaxation times of 4.5 ms and 0.1 s; at 5 mg/ml enzyme the corresponding values are 2.8 ms and 31 ms. Assuming the simplest mechanism consistent with these results, estimate the values of the rate constants.

9.2 The following data can be described by an equation of the form $y = A + B \exp(-t/\tau_1) + C \exp(-t/\tau_2)$. Assuming that τ_1 is less than 5 ms, estimate the values of A, C and τ_2, and use the results to estimate B and τ_1. Values of t are in milliseconds.

t	y	t	y	t	y
1	78	7	40	25	20
2	68	8	37	30	19
3	60	9	34	35	18
4	53	10	32	40	17
5	48	15	25	45	17
6	43	20	22	50	17

9.3 In steady-state studies of an enzyme, a competitive inhibitor was found to have $K_i = 20 \, \mu$M. This K_i value was interpreted as a true equilibrium constant. The Michaelis constant for the substrate was found to be 0.5 mM. Subsequently the following stopped-flow experiments were carried out: in experiment (a), one syringe of the apparatus contained 0.2 mM inhibitor, and the other contained 50 μM enzyme, 0.2 mM inhibitor and 10 mM substrate; in experiment (b), one syringe contained 0.2 mM inhibitor and 50 μM enzyme, and the other contained 0.2 mM inhibitor and 10 mM substrate. In experiment (a), the transient phase of the reaction had a time constant of 7.7 ms, and in experiment (b) the time constant was 15 ms. Estimate the *on* and *off* rate constants for binding of the inhibitor to the free enzyme. Do the results of the stopped-flow experiments require the interpretation of K_i as an equilibrium constant to be revised?

Chapter 10

Estimation of kinetic constants

10.1 Cautionary note

In this chapter I shall outline the statistical methods that can be used to estimate kinetic constants from experimental data. However, to avoid devoting an inordinate amount of space to a discussion of the underlying principles I shall make only passing reference to the assumptions implicit in statistical methods. Instead, I shall assume that the error structure that has most often been observed in enzyme kinetics (Storer, Darlison and Cornish-Bowden, 1975; Siano, Zyskind and Fromm, 1975; Askelöf, Korsfeldt and Mannervik, 1976) is the one that usually occurs. However, there may well be exceptions in practice and so one should always be aware of the possibility that the weights used in this (or any other) textbook may be incorrect. This may seem a tedious and fussy point that one can safely ignore, but there is no advantage, and there may be a serious disadvantage, in using a wrongly weighted statistical method; one may as well — indeed better — fit lines by eye. It is not difficult to find statements in the literature that make it quite clear that the authors do not understand the method they are using: for example 'all data points [on a double-reciprocal plot] were given equal weight after they fell within a 5% error limit during a test of linearity'. (This is a genuine quotation from a paper published in a reputable journal in 1976. The misconception that it embodies, however, is so widespread that it would hardly be fair to single out the authors by providing a reference.) The observation in this case implies a constant coefficient of variation, or constant standard deviation expressed as a percentage, and not a constant variance as implied by equal weighting.

Most of the methods described in this chapter are suitable for expression as computer programs — indeed, numerous programs are available to enzymologists from various sources: these have been critically reviewed by Garfinkel, Kohn and Garfinkel (1977). Such programs are very valuable and the serious kineticist nowadays would find it difficult to proceed without use of computers. Nonetheless the warning given above applies with even greater force to analysis of data by computer. All data-fitting programs either incorporate some assumptions about the nature of the experimental error or they require the user to choose between various sets of assumptions. In both cases, some understanding on the part of the user is essential. Blind use of a program written by someone else is rarely a safe procedure. It cannot convert a badly designed experiment into a well designed experiment and it cannot convert poor data into precise information. In the worst case, if the program incorporates assumptions that are completely inappropriate to the experiment, it may provide much less accurate information than one could obtain by inspecting a graph.

10.2 Least-squares fit to the Michaelis–Menten equation

The Michaelis–Menten equation as it is usually written (e.g. equation 2.7) is incomplete, because it ignores the effect of experimental error. A better way of writing it is

$$v_i = \frac{Vs_i}{K_m + s_i} (1 + e_i) \qquad (10.1)$$

The subscripts i are to show that we shall be dealing, not with a single isolated observation, but with the ith of a *sample* of n observations. The error term $(1 + e_i)$ appears as a *factor* in the equation, rather than a simple additive error, because the few investigations of experimental error in enzyme kinetics (Storer, Darlison and Cornish-Bowden, 1975; Siano, Zyskind and Fromm, 1975; Askelöf, Korsfeldt and Mannervik, 1976) suggest that it is commonly a better approximation to assume that rate measurements are subject to a constant *coefficient of variation* than to assume a constant *standard deviation*. Thus e_i measures the *relative* deviation of the observed rate v_i from the calculated rate $Vs_i/(K_m + s_i)$, not the simple difference between the observed and calculated rates. The term *deviation* is appropriate rather than *error*, because it is measured in relation to the calculated rate, not the true rate, which is unknown. In general, the true values of V and K_m can never be known regardless of how many measurements may be made; the best one can hope for is to estimate them accurately.

When the Michaelis–Menten equation is expressed as equation 10.1, it is easy to see why linear transformations such as those in equations 2.10–2.12 (Section 2.5) can lead to unsatisfactory results. As they were obtained from equation 2.7 by perfectly correct algebra it is not obvious at first how they can be faulty. But once it is realized that it is not the algebra but the starting point that is invalid, and that the complete expression, equation 10.1, cannot be recast as the equation for a straight line, the difficulty disappears.

The problem of estimating V and K_m as accurately as possible may be expressed as one of finding the values that make the deviations e_i as small as possible. Except in the trivial case where there are not more than two observations, it is not in general possible to find values that make all the e_i zero. Instead, one can make the average of all the e_i^2 as small as possible. We use e_i^2 here rather than e_i to avoid complications due to the occurrence of both positive and negative deviations (In principle one might achieve a similar effect by simply ignoring the signs of the e_i, but this leads to such difficult algebra that it is not usually done.) Thus we can define the best-fit values of V and K_m as those that jointly minimize the *sum of squares* of deviations, *SS*, defined as

$$SS = \sum_{i=1}^{n} e_i^2$$

By rearranging equation 10.1, we may express e_i in terms of v_i, s_i, K_m and V:

$$e_i = \frac{K_m v_i}{V s_i} + \frac{v_i}{V} - 1$$

and hence

$$SS = \Sigma (a v_i / s_i + b v_i - 1)^2$$

where $a = K_m / V$ and $b = 1/V$ are the intercept on the ordinate and slope respectively of a plot of s_i / v_i against s_i (cf. Section 2.5 and *Figure 2.4*). In principle, we can find the values of V and K_m that make *SS* a minimum by partially differentiating with respect to V and K_m and setting both derivatives to zero. But we can reach the same conclusion by a shorter and simpler route by solving for a and b first. Partial differentiation gives

$$\frac{\partial SS}{\partial a} = \Sigma \left[\frac{2 v_i}{s_i} \left(\frac{a v_i}{s_i} + b v_i - 1 \right) \right]$$

$$\frac{\partial SS}{\partial b} = \Sigma \left[2 v_i \left(\frac{a v_i}{s_i} + b v_i - 1 \right) \right]$$

If we define $\hat{a}$ and $\hat{b}$ as the values of a and b respectively that make *SS* a minimum, we can set both expressions to zero and rearrange them to give a pair of simultaneous equations in $\hat{a}$ and $\hat{b}$:

$$\hat{a} \Sigma v_i^2 / s_i^2 + \hat{b} \Sigma v_i^2 / s_i = \Sigma v_i / s_i$$

$$\hat{a} \Sigma v_i^2 / s_i + \hat{b} \Sigma v_i^2 = \Sigma v_i$$

which may be solved to give

$$\hat{a} = \frac{\Sigma v_i^2 \ \Sigma v_i/s_i - \Sigma v_i^2/s_i \ \Sigma v_i}{\Sigma v_i^2/s_i^2 \ \Sigma v_i^2 - (\Sigma v_i^2/s_i)^2}$$

$$\hat{b} = \frac{\Sigma v_i^2/s_i^2 \ \Sigma v_i - \Sigma v_i^2/s_i \ \Sigma v_i/s_i}{\Sigma v_i^2/s_i^2 \ \Sigma v_i^2 - (\Sigma v_i^2/s_i)^2}$$

These lead straightforwardly to expressions for the best-fit values $\hat{V}$ and $\hat{K}_m$ of V and K_m respectively:

$$\hat{V} = 1/\hat{b} = \frac{\Sigma v_i^2/s_i^2 \ \Sigma v_i^2 - (\Sigma v_i^2/s_i)^2}{\Sigma v_i^2/s_i^2 \ \Sigma v_i - \Sigma v_i^2/s_i \ \Sigma v_i/s_i}$$

$$\hat{K}_m = \hat{a}/\hat{b} = \frac{\Sigma v_i^2 \ \Sigma v_i/s_i - \Sigma v_i^2/s_i \ \Sigma v_i}{\Sigma v_i^2/s_i^2 \ \Sigma v_i - \Sigma v_i^2/s_i \ \Sigma v_i/s_i}$$

This result was first given by Johansen and Lumry (1961), and is exact; no further refinement is necessary to minimize SS. With other assumptions about the distribution of experimental error, for example that each rate has the same variance, the fitting problem is somewhat more difficult, because the equations that define the best-fit solution have no analytical solution. Consequently one must first obtain an approximate solution and then refine it by means of successively better approximations. Johansen and Lumry (1961), and also Wilkinson (1961), describe how this can be done. *See also* Problem 10.3.

10.3 Statistical aspects of the direct linear plot

The least-squares approach to data-fitting problems is the most convenient general method, but it is not necessarily the best. To demonstrate that the least-squares solution to a problem is the 'best' solution, one must assume (a) that the random errors in the measurement are distributed according to the normal curve of error; (b) that only one measured variable, the so-called *dependent variable*, is subject to experimental error; (c) that the correct weights are known; (d) that the errors are uncorrelated, that is, that the magnitude of a particular error implies nothing about the magnitude of any other error; and (e) that systematic error can be ignored, that is, that the distribution curve for each error has a mean value of zero. The need for most of these assumptions is generally agreed, but the first, the assumption of a normal distribution, requires some comment, because it is sometimes asserted that the least-squares method retains some of its optimal properties for all distributions. For example, Garfinkel, Kohn and Garfinkel (1977), in an otherwise authoritative review, state that least-squares estimators are 'minimum-variance' estimators for any distribution of errors. However, reference to the source quoted for this statement (Kendall and Stuart, 1973) shows that it is true only if one restricts consideration to a particular class of estimators known as *linear estimators*. It is not true in general, and under some circumstances the estimators derived from the direct linear plot, which are not linear estimators, have smaller variances (i.e. they are more precise) than the best least-squares estimators.

Unfortunately one knows little in practice about the truth of any of the assumptions implicit in least-squares estimation. Most scientists prefer their conclusions to depend as little as possible on unproved assumptions, and a different branch of statistics, known as *distribution-free* or *non-parametric*, dispenses with all of the above assumptions except the last. The last is retained but in a weaker form: one assumes in the absence of other information that the error in any measurement is as likely to be positive as to be negative.

The direct linear plot of Eisenthal and Cornish-Bowden (1974) (*see* Section 2.5) represents an attempt at introducing distribution-free ideas into enzyme kinetics, at the same time greatly simplifying the procedures and concepts. For any non-duplicate pair of observations (s_i, v_i) and (s_j, v_j), there is a unique pair of values of V and K_m in the Michaelis–Menten equation that satisfy both observations exactly:

$$V_{ij} = \frac{s_i - s_j}{(s_i/v_i) - (s_j/v_j)}$$

$$K_{ij} = \frac{v_j - v_i}{(v_i/s_i) - (v_j/s_j)}$$

These values define the co-ordinates of the point of intersection of the lines drawn for the two observations as described in Section 2.5. Altogether, n observations provide a maximum of $\frac{1}{2}n(n - 1)$ such pairs of values. (This is the maximum number rather than the actual number because the pairs of values obtained from duplicate observations with $s_i = s_j$ are meaningless and must be omitted from consideration.) The *median* of the set of K_{ij} values can be defined as K_m^*, the best-fit estimate of K_m, and the median of the set of V_{ij} values as V^*, the corresponding best-fit estimate of V. The median is the middle one of a set of values arranged in rank order, or the mean of the middle pair if the number of values is even. There are two main reasons for using the median in the present context rather than any other sort of average, such as the more familiar arithmetic mean: first, the estimates K_{ij} and V_{ij} are automatically ranked by the direct linear plot, as illustrated in *Figure 10.1*, and no calculation is required to find their

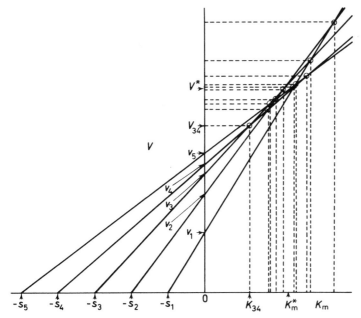

Figure 10.1 Determination of median estimates from the direct linear plot. The lines are drawn as in *Figure 2.6*, and each intersection (shown as a circle) provides an estimate K_{ij} of K_m, and an estimate V_{ij} of V. These estimates are marked off on the axes for clarity. The best-fit estimate K_m^* is then taken as the median (i.e. the middle) K_{ij} value, and V^* as the median V_{ij} value. If there are an even number of intersections, as in this example, the median is taken as the mean of the middle two values. If there are replicate observations (i.e. measurements at the same value of s), the intersections on the K_m axis that these generate must not be included in determining the medians. Intersections in the second quadrant (negative K_{ij}, positive V_{ij}) are taken at face value; intersections in the third quadrant (negative K_{ij} and negative V_{ij}) are interpreted as giving large *positive* values of *both* K_{ij} and V_{ij}. If intersections outside the first quadrant are numerous, or if the arrangement of intersections shows a clear and reproducible pattern, the possibility should be examined that the data do not fit the Michaelis–Menten equation

medians; second, the median, unlike other types of average, is insensitive to extreme values, which inevitably occur sometimes in the direct linear plot, because some of the lines are nearly parallel.

The main advantage of this approach over the method of least squares is that it requires no calculation, and it incorporates no abstract or difficult idea such as the normal distribution of errors. However, it also has some practical statistical advantages (Cornish-Bowden and Eisenthal, 1974): it is only marginally inferior to the method of least squares even when the least-squares assumptions are correct, and in other circumstances it may be much superior, because it incorporates few and weak assumptions and is consequently insensitive to departures from the expected error structure. For example, if one calculates least-squares estimates with the assumption of a constant coefficient of variation in the observed rates, as in Section 10.2, when in reality the rates have constant variance, or vice versa, the results will be much less precise than those given by the distribution-free method. Alternatively, if the experiment contains an undetected 'wild observation', or *outlier*, this will have a drastic and damaging effect on the least-squares calculation but very little effect on the distribution-free calculation. Under ideal conditions the least-squares method will always perform a little better, but it is very sensitive to departures from ideal conditions, and consequently it often performs much worse.

In experiments with rather large errors in the rates, the direct linear plot as originally described has a slight negative bias, i.e. it

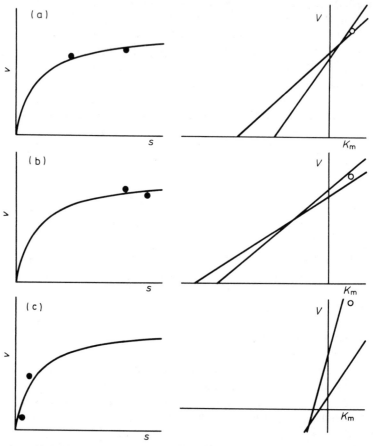

Figure 10.2 Explanation of why intersections may occur in the second or third quadrants of the direct linear plot. In each row, the left side shows two points on a plot of v against s, and the right side shows the corresponding direct linear plot, with an open circle (o) to indicate the values of K_m and V used in calculating curves shown on the left. (a) The 'normal' case, with an intersection in the first quadrant. (b) Intersections in the second quadrant occur typically when both s values are large compared with K_m, and both v values are similar in magnitude to V, but are incorrectly ranked. (c) Intersections in the third quadrant occur typically when one (and often both) of the s values is small compared with K_m, and at least one of the v values is small compared with V

leads to V^* and K_m^* values that are not distributed about the corresponding true values but about values that are too small (Cornish-Bowden and Eisenthal, 1978). This bias derives from the occurrence of intersections in the third quadrant, with negative values of both V_{ij} and K_{ij}, and to understand how to correct the bias it is useful to consider why intersections should ever occur outside the first quadrant. *Figure 10.2* shows typical combinations of s and v values that lead to intersections in the first, second and third quadrants. (In all cases I assume that the anomalies arise from statistical error and not from failure to fit the right equation: this may be reasonable if there are only a small minority of anomalous intersections, but it would not be reasonable if there are many or if the whole arrangement of intersections in the plot appears systematic rather than chaotic.) It will be seen that intersections in the second and third quadrants occur for quite different reasons; they consequently require quite different treatment. Intersections in the second quadrant (*Figure 10.2b*) typically occur when both s_i and s_j are large compared with K_m, so that v_i and v_j are similar in magnitude to V, but there is not enough information to assign a value to K_m except that it is small compared with both s_i and s_j. Thus if experimental error causes the two v values to be incorrectly ranked, as in the example in *Figure 10.2b*, it is nonetheless reasonable to take both K_{ij} and V_{ij} at face value and accord them no special treatment in finding the medians K_m^* and V^*. Intersections in the third quadrant (*Figure 10.2c*), on the other hand, typically occur when s_i and s_j are both so much smaller than K_m that there is not enough information to assign values to either K_m or V except that they are *large* compared with the s and v values. In this case, if experimental error causes the two v/s values to be incorrectly ranked, so that the corresponding K_{ij} and V_{ij} values are *both* negative, as in the example in *Figure 10.2c*, treatment of these values at face value in finding K_m^* and V^* will plainly result in negative bias: this difficulty can simply and logically be overcome by treating *both* K_{ij} and V_{ij} as if they had very large *positive* values. In determining the medians K_m^* and V^* one does not require numerical values of the extreme members of the samples. An alternative plot, which avoids the need for special rules, is described by Cornish-Bowden and Eisenthal (1978).

10.4 Precision of K_m and V estimates

I have tried to keep mathematical complexities to a minimum in this chapter, and consequently I have not discussed how to estimate the precision of K_m and V or other fitted parameters. Instead, this section will provide a brief guide to other sources of information. Johansen and Lumry (1961) and Wilkinson (1961) both give full accounts of the method of least squares as applied to the Michaelis–Menten equation, and provide some information about the treatment of other problems of importance in enzyme kinetics. Cornish-Bowden and Eisenthal (1974) describe a distribution-free method of finding a joint confidence region for K_m and V, but not for either parameter considered by itself; but Porter and Trager (1977) have remedied this omission. Cornish-Bowden, Porter and Trager (1978) examined the reliability of both least-squares and distribution-free confidence limits. They found, not surprisingly, that least-squares confidence limits were satisfactory when correctly weighted but could be seriously misleading otherwise; the distribution-free method proved more robust.

Several reviews of statistical methods in enzyme kinetics exist and may be consulted for further information, such as those of Reich (1970), Markus *et al.* (1976) and Garfinkel, Kohn and Garfinkel (1977).

10.5 Examination of residuals

After data have been fitted to an equation, it is always worth while to check whether the assumptions made in the analysis were reasonable. Has one fitted the right equation, or would a more complex one have given a more reasonable explanation of the observations? Has one made reasonable statistical assumptions — for example, is the coefficient of variation a constant, as assumed in Section 10.2, or would a constant variance be a better approximation, or does the experimental error vary in a more complex way than either?

Much the easiest way of attacking these questions is to examine the *residuals* after fitting, that is, the differences between observed rates v and the corresponding rates $\hat{v}$ calculated from the best-fit equation. I shall not attempt an exhaustive treatment in this section, but will rather indicate the sort of uses to which residuals can be put and mention some sources of further information.

If the assumptions of Section 10.2 are correct, and the correct equation is fitted, the simple differences $(v - \hat{v})$ should tend to increase in absolute magnitude as the calculated rate $\hat{v}$ increases, but the *relative* differences $(v - \hat{v})/\hat{v}$ should be scattered in a parallel band about zero. On the other hand, if the assumptions are incorrect, and the observed rates actually have a constant variance, the simple differences should be scattered in a parallel band about zero whereas the relative differences should tend to decrease in absolute magnitude as $\hat{v}$ increases. *Figure 10.3* shows examples of such plots. They are easy to execute, regardless of the complexity of the equation that has been fitted, and may provide valuable information that is not readily available in any other way. They do, however, require a

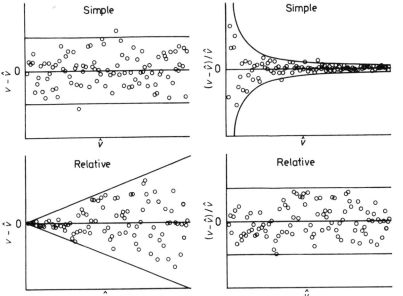

Figure 10.3 Scatter plots for assessing the correctness of a weighting scheme. The upper pair of plots shows the expected results of plotting the deviation from the fitted line, $v - \hat{v}$, against the rate, $\hat{v}$, and of plotting $(v - \hat{v})/\hat{v}$ against $\hat{v}$, in the event that all the rates have the same standard deviation ('simple errors'). The lower pair of plots shows the expected results in the event that the standard deviation of each rate is proportional to its true value ('relative errors'). In each plot, limits are drawn for deviations of twice the standard deviation. The least-squares expressions given in this chapter were derived assuming 'relative errors'

large number of observations (of the order of 50 or more), because the scatter is always so large that consideration of only a few points may give a completely misleading impression.

Sometimes a plot of residuals against $\hat{v}$ may give a completely different appearance from any of those in *Figure 10.3*, for example the one shown in *Figure 10.4a*, which is taken from a study of butyrylcholinesterase by Augustinsson, Bártfai and Mannervik (1974). It is immediately evident that there is relatively little scatter, and any-way the points are scattered not about zero but about a well-defined curve. This sort of result provides a clear indication that the data have been fitted to an unsuitable equation and that a different one, pro-bably a more complex one, is required. Although one might in prin-ciple make the same sort of deduction from a more conventional plot of v against s (or one of the common straight-line plots), by noting systematic deviations of the points from the calculated line, or by noting a systematic pattern of + and − signs in the column of residuals in a computer printout, the appearance of the residual plot is unmistakable. When the data used for *Figure 10.4a* were fitted to a more suitable equation, the resulting residual plot (*Figure 10.4b*) had a much more random appearance similar to that of *Figure 10.3a*,

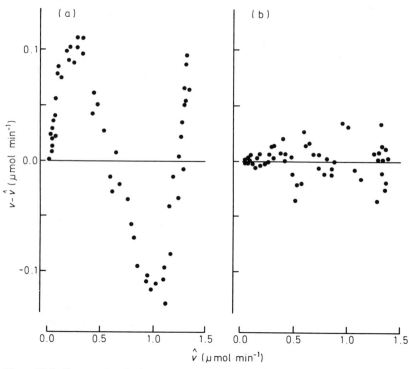

Figure 10.4   Scatter plots for butyrylcholinesterase (Augustinsson, Bártfai and Mannervik, 1974). The points shown in (a) were obtained after fitting kinetic data to the Michaelis–Menten equation, and show a clear systematic trend, which became a random scatter (b) when the fitting procedure was repeated with a more suitable rate equation.

and suggested that Augustinsson, Bártfai and Mannervik (1974) were correct in assuming a constant coefficient of variation.

If a residual plot does indicate that a more complex equation is required, the question is then to decide what more complex equation to try. Again, residual plots should be helpful. Suppose, for example, that one is studying the effect of an inhibitor and that one has fitted the data assuming the inhibition to be competitive when in fact there is a small but appreciable uncompetitive component. In this case the approximation may well give good results at low substrate concentrations but very poor results at high substrate concentrations, where competitive effects become less important but uncompetitive effects become more important. So a plot of residuals against substrate concentrations (with points for all inhibitor concentrations included on the same plot) should show a definite systematic trend that would not be evident from a plot of the same residuals against the calculated rate.

For further information about plots of residuals, papers by Mannervik and Bártfai (1973) and by Ellis and Duggleby (1978) should be consulted.

Problems

10.1 Determine least-squares and distribution-free estimates of the parameters of the Michaelis–Menten equation for the following set of data:

s (mM)	v (mM min^{-1})	s (mM)	v (mM min^{-1})
1	0.219	6	0.525
2	0.343	7	0.512
3	0.411	8	0.535
4	0.470	9	0.525
5	0.490	10	0.540

For both sets of estimates, plot residuals against s. Is any trend apparent? If so, what experiments would you carry out to decide whether the trend is real and not an artefact of random error? Can any conclusions be drawn about the weights appropriate for least-squares analysis?

10.2 What would be the parameter estimates for the data in Problem 10.1 if the value of v for s = 1 mM were 0.159 mM min^{-1} instead of 0.219 mM min^{-1}? Discuss.

10.3 If all of the v_i values in an experiment were thought to have equal variance, it would be appropriate to replace equation 10.1 by the following equation:

$$v_i = \frac{V s_i}{K_m + s_i} + e_i$$

and in this case SS would be defined as follows:

$$SS = \Sigma e_i^2 = \Sigma w_i (a + b s_i - s_i/v_i)^2$$

where $a = K_m/V$, $b = 1/V$ and each w_i is a *weight* approximately equal to v_i^4/s_i^2. Treating this approximation as if it were exact, derive revised expressions for $\hat{K}_m$ and $\hat{V}$ that minimize the revised expression for SS.

10.4 Check your solution to Problem 10.3 for dimensional consistency.

References

The sections of the book where the work is referred to are given in brackets after each reference.

G.S. ADAIR (1925a) *J. biol. Chem.* **63**, 529–545. [8.4]

G.S. ADAIR (1925b) *Proc. Roy. Soc., Ser. A* **109**, 292–300. [8.4]

R.A. ALBERTY (1953) *J. Amer. chem. Soc.* **75**, 1928–1932. [6.3]

R.A. ALBERTY (1958) *J. Amer. chem. Soc.* **80**, 1777–1782. [6.3]

W.J. ALBERY and J.R. KNOWLES (1976) *Biochemistry* **15**, 5627–5631. [6.8]

S. ARRHENIUS (1889) *Z. physik. Chem.* **4**, 226–248. [1.6]

P. ASKELÖF, M. KORSFELDT and B. MANNERVIK (1976) *Eur. J. Biochem.* **69**, 61–67. [10.1, 10.2]

D.E. ATKINSON (1977) *Cellular Energy Metabolism and its Regulation* Academic Press, New York. [8.1]

K.-B. AUGUSTINSSON, T. BÅRTFAI and B. MANNERVIK (1974) *Biochem. J.* **141**, 825–834. [10.5]

M.L. BENDER, M.L. BEGUÉ-CANTÓN, R.L. BLAKELEY, L.J. BRUBACHER, J. FEDER, C.R. GUNTER, F.J. KÉZDY, J.V. KILLHEFFER, JR., T.H. MARSHALL, C.G. MILLER, R.W. ROESKE and J.K. STOOPS (1966) *J. Amer. chem. Soc.* **88**, 5890–5913. [9.2]

M.L. BENDER, F.J. KÉZDY and C.R. GUNTER (1964) *J. Amer. chem. Soc.* **86**, 3714–3721. [7.7]

D. BLANGY, H. BUC and J. MONOD (1968) *J. mol. Biol.* **31**, 13–35. [8.7]

V. BLOOMFIELD, L. PELLER and R.A. ALBERTY (1962) *J. Amer. chem. Soc.* **84**, 4367–4374. [6.3]

C. BOHR (1903) *Zentralbl. Physiol.* **17**, 682–688. [8.2]

J. BOTTS and M. MORALES (1953) *Trans. Faraday Soc.* **49**, 696–707. [4.2, 4.3, 4.4, 5.8]

P.D. BOYER (1959) *Arch. Biochem. Biophys.* **82**, 387–410. [6.8]

M.M. BRADFORD (1976) *Analyt. Biochem.* **72**, 248–254. [3.4]

G.E. BRIGGS and J.B.S. HALDANE (1925) *Biochem. J.* **19**, 338–339. [2.3]

R. BRINKMAN, R. MARGARIA and F.J.W. ROUGHTON (1933) *Phil. Trans. Roy. Soc. Lond., Ser. A* **232**, 65–97. [9.3]

J.N. BRØNSTED (1923) *Rec. Trav. Chim. Pays-Bas* **42**, 718–728. [7.2]

A.J. BROWN (1892) *J. chem. Soc. (Trans.)* **61**, 369–385. [2.1]

A.J. BROWN (1902) *J. chem. Soc. (Trans.)* **81**, 373–388. [2.1]

E. BUCHNER (1897) *Ber. dt. chem. Ges.* **30**, 117–124. [2.1]

J.J. BURKE, G.G. HAMMES and T.B. LEWIS (1965) *J. chem. Phys.* **42**, 3520–3525. [9.4]

M.L. CARDENAS, E. RABAJILLE and H. NIEMEYER (1978) *Arch. Biochem. Biophys.* **190**, 142–148. [8.10]

H. CEDAR and J.H. SCHWARTZ (1969) *J. biol. Chem.* **244**, 4122–4127. [6.8]

S. CHA (1968) *J. biol. Chem.* **243**, 820–825. [4.3]

B. CHANCE (1940) *J. Franklin Inst.* **229**, 455–476, 613–640, 737–766. [9.3]

B. CHANCE (1951) *Advan. Enzymol.* **12**, 153–190. [9.3]

R.E. CHILDS and W.G. BARDSLEY (1975) *J. theor. Biol.* **53**, 381–394. [Problem 5.1]

K. CHOU, S. JIANG, W. LIU and C. FEE (1979) *Sci. Sin.* **22**, 341–358. [4.2]

W.W. CLELAND (1963) *Biochim. biophys. Acta* **67**, 104–137. [6.3]

COMMISSION ON BIOCHEMICAL NOMENCLATURE (1973) *Recommendations (1972) of the Commission on Biochemical Nomenclature on the Nomenclature and Classification of Enzymes together with their Units and the Symbols of Enzyme Kinetics* Elsevier, Amsterdam. [2.3, 6.3]

A. CONWAY and D.E. KOSHLAND, JR. (1968) *Biochemistry* **7**, 4011–4023. [8.8]

R.A. COOK and D.E. KOSHLAND, JR. (1970) *Biochemistry* **9**, 3337–3342. [8.5]

A.J. CORNISH-BOWDEN (1967) *D. Phil. Thesis* University of Oxford. [5.5]

A. CORNISH-BOWDEN (1974) *Biochem. J.* **137**, 143–144. [5.5]

A. CORNISH-BOWDEN (1975) *Biochem. J.* **149**, 305–312. [2.8, 3.2]

A. CORNISH-BOWDEN (1976) *Biochem. J.* **153**, 455–461. [2.6, 7.6]

A. CORNISH-BOWDEN (1979) *Eur. J. Biochem.* **93**, 383–385. [Problem 5.1]

A. CORNISH-BOWDEN and R. EISENTHAL (1974) *Biochem. J.* **139**, 721–730. [10.3]

A. CORNISH-BOWDEN and R. EISENTHAL (1978) *Biochim. biophys. Acta* **523**, 268–272. [10.3]

A.J. CORNISH-BOWDEN and J.R. KNOWLES (1969) *Biochem. J.* **113**, 353–362. [7.5]

A. CORNISH-BOWDEN and D.E. KOSHLAND, JR. (1970) *Biochemistry* **9**, 3325–3336. [8.5]

A. CORNISH-BOWDEN and D.E. KOSHLAND, JR. (1975) *J. mol. Biol.* **95**, 201–212. [8.5]

A. CORNISH-BOWDEN, W.R. PORTER and W.F. TRAGER (1978) *J. theor. Biol.* **74**, 163–175. [10.4]

A. CORNISH-BOWDEN and J.T. WONG (1978) *Biochem. J.* **175**, 969–976. [6.3]

K. DALZIEL (1957) *Acta Chem. Scand.* **11**, 1706–1723. [6.3, Problem 6.8]

K. DALZIEL (1969) *Biochem. J.* **114**, 547–556. [6.9]

M.P. DEUTSCHER (1967) *J. biol. Chem.* **242**, 1123–1131. [3.6]

H.B.F. DIXON (1973) *Biochem. J.* **131**, 149–154. [7.6]

H.B.F. DIXON (1976) *Biochem. J.* **153**, 627–629. [7.3]

H.B.F. DIXON (1979) *Biochem. J.* **177**, 249–250. [7.3]

M. DIXON (1953a) *Biochem. J.* **55**, 161–170. [7.4]

M. DIXON (1953b) *Biochem. J.* **55**, 170–171. [5.5]

M. DIXON and E.C. WEBB (1963) *Enzymes* 2nd edn, Longmans, London, pp. 296–304. [6.9]

M. DOUDOROFF, H.A. BARKER and W.Z. HASSID (1947) *J. biol. Chem.* **168**, 725–732. [6.2, 6.8]

C. DOUMENG and S. MAROUX (1979) *Biochem. J.* **177**, 801–808. [Problem 5.7]

G.S. EADIE (1942) *J. biol. Chem.* **146**, 85–93. [2.5]

R.R. EADY, D.J. LOWE and R.N.F. THORNELEY (1978) *FEBS Lett.* **95**, 211–213. [9.3]

M. EIGEN (1954) *Discuss. Faraday Soc.* **17**, 194–205. [9.4]

R. EISENTHAL and A. CORNISH-BOWDEN (1974) *Biochem. J.* **139**, 715–720. [2.5, 10.3]

K.R.F. ELLIOTT and K.F. TIPTON (1974) *Biochem. J.* **141**, 789–805. [6.9]

K.J. ELLIS and R.G. DUGGLEBY (1978) *Biochem. J.* **171**, 513–517. [10.5]

A. ESEN (1978) *Analyt. Biochem.* **89**, 264–273. [3.4]

H. EYRING (1935) *J. chem Phys.* **3**, 107–115. [1.7]

W. FERDINAND (1966) *Biochem. J.* **98**, 278–283. [8.10]

A.R. FERSHT (1977) *Enzyme Structure and Mechanism* Freeman, Reading and San Francisco, pp. 137–140 [Problem 7.1]; pp. 129–132 [9.4]

A.R. FERSHT and R. JAKES (1975) *Biochemistry* **14**, 3350–3356. [9.3]

E. FISCHER (1894) *Ber. dt. chem. Ges.* **27**, 2985–2993. [8.6]

J.R. FISHER and V.D. HOAGLAND, JR. (1968) *Adv. biol. med. Phys.* **12**, 163–211. [6.2]

C. FRIEDEN (1959) *J. biol. Chem.* **234**, 2891–2896. [6.9]

C. FRIEDEN (1967) *J. biol. Chem.* **242**, 4045–4052. [8.9]

C. FRIEDEN and R.F. COLMAN (1967) *J. biol. Chem.* **242**, 1705–1715. [8.9]

L. GARFINKEL, M.C. KOHN and D. GARFINKEL (1977) *CRC Crit. Rev. Bioengng* **2**, 329–361. [10.1, 10.3, 10.4]

Q.H. GIBSON and L. MILNES (1964) *Biochem. J.* **91**, 161–171. [9.3]

E.A. GUGGENHEIM (1926) *Phil. Mag. Ser. VII* **2**, 538–543. [1.5]

J.S. GULBINSKY and W.W. CLELAND (1968) *Biochemistry* **7**, 566–575. [4.3, 6.2]

H. GUTFREUND (1955) *Discuss. Faraday Soc.* **20**, 167–173. [9.2]

J.E. HABER and D.E. KOSHLAND, JR. (1967) *Proc. natn. Acad. Sci. U.S.* **58**, 2087–2093. [8.8]

J.B.S. HALDANE (1930) *Enzymes* Longmans Green, London. [2.6, 3.6, 6.2]

G.G. HAMMES and P. FASELLA (1962) *J. Amer. chem. Soc.* **84**, 4644–4650. [9.6]

G.G. HAMMES and P.R. SCHIMMEL (1970) in P. Boyer (Editor), *The Enzymes* 3rd edn, Academic Press, New York, Vol. 2, pp. 67–114. [9.4, 9.7]

C.S. HANES (1932) *Biochem. J.* **26**, 1406–1421. [2.5]

C.S. HANES, P.M. BRONSKILL, P.A. GURR and J.T. WONG (1972) *Can. J. Biochem.* **50**, 1385–1413. [6.2]

A.V. HARCOURT (1867) *J. chem. Soc.* **20**, 460–492. [1.6]

B.S. HARTLEY and B.A. KILBY (1954) *Biochem. J.* **56**, 288–297. [9.2, 9.6]

H. HARTRIDGE and F.J.W. ROUGHTON (1923) *Proc. Roy. Soc., Ser. A* **104**, 376–394, 395–430. [9.3]

J.W. HASTINGS and Q.H. GIBSON (1963) *J. biol. Chem.* **238**, 2537–2554. [9.3]

V. HENRI (1902) *C. r. hebd. Acad. Sci., Paris* **135**, 916–919. [2.2]

V. HENRI (1903) *Lois Générales de l'Action des Diastases* Hermann, Paris. [2.2]

A.V. HILL (1910) *J. Physiol.* **40**, iv–vii. [8.2, 8.3]

C.M. HILL, R.D. WAIGHT and W.G. BARDSLEY (1977) *Mol. cell. Biochem.* **15**, 173–178. [3.7]

D.I. HITCHCOCK (1926) *J. Amer. chem. Soc.* **48**, 2870. [2.2]

B.H.J. HOFSTEE (1952) *J. biol. Chem.* **199**, 357–364. [2.5]

M.R. HOLLAWAY and H.A. WHITE (1975) *Biochem. J.* **149**, 221–231. [9.3]

M.J. HOLROYDE, M.B. ALLEN, A.C. STORER, A.S. WARSY, J.M.E. CHESHER, I.P. TRAYER, A. CORNISH-BOWDEN and D.G. WALKER (1976) *Biochem. J.* **153**, 363–373. [3.5, 8.10]

C.S. HUDSON (1908) *J. Amer. chem. Soc.* **30**, 1564–1583. [3.6]

A. HUNTER and C.E. DOWNS (1945) *J. biol. Chem.* **157**, 427–446. [5.3]

D.W. INGLES and J.R. KNOWLES (1967) *Biochem. J.* **104**, 369–377. [5.10]

W.P. JENCKS (1969) *Catalysis in Chemistry and Enzymology* McGraw-Hill, New York, pp. 471–477, 590–597. [7.6]

R.R. JENNINGS and C. NIEMANN (1955) *J. Amer. chem. Soc.* **77**, 5432–5483. [2.8]

G. JOHANSEN and R. LUMRY (1961) *C. r. trav. Lab. Carlsberg* **32**, 185–214. [10.2, 10.4]

F.C. KAFATOS, A.M. TARTAKOFF and J.H. LAW (1967) *J. biol. Chem.* **242**, 1477–1487. [3.1]

M.G. KENDALL and A. STUART (1973) *The Advanced Theory of Statistics* 3rd edn, Griffin, London, Vol. 2. [10.3]

J.C. KENDREW, R.E. DICKERSON, B.E. STRANDBERG, R.G. HART, D.R. DAVIES, D.C. PHILLIPS and V.C. SHORE (1960) *Nature, Lond.* **185**, 422–427. [8.2]

E.L. KING and C. ALTMAN (1956) *J. phys. Chem.* **60**, 1375–1378. [4.1, 4.2, 6.2]

M.E. KIRTLEY and D.E. KOSHLAND, JR. (1967) *J. biol. Chem.* **242**, 4192–4205. [8.8]

R. KITZ and I.B. WILSON (1962) *J. biol. Chem.* **237**, 3245–3249. [5.1, Problem 5.1]

J.R. KNOWLES (1976) *CRC Crit. Rev. Biochem.* **4**, 165–173. [7.6]

J.R. KNOWLES, R.S. BAYLISS, A.J. CORNISH-BOWDEN, P. GREENWELL, T.M. KITSON, H.C. SHARP and G.B. WYBRANDT (1970) in P. Desnuelle, H. Neurath and M. Ottesen (Editors), *Structure–Function Relationships of Proteolytic Enzymes* Munksgaard, Copenhagen, pp. 237–250. [7.2]

D.E. KOSHLAND, JR. (1954) in W.D. McElroy and B. Glass (Editors), *A Symposium on the Mechanism of Enzyme Action* Johns Hopkins Press, Baltimore, pp. 608–641. [6.2]

D.E. KOSHLAND, JR. (1955) *Discuss. Faraday Soc.* **20**, 142–148. [6.8]

D.E. KOSHLAND, JR. (1958) *Proc. natn. Acad. Sci. U.S.* **44**, 98–99. [6.2, 8.6]

D.E. KOSHLAND, JR. (1959a) in P.D. Boyer, H. Lardy and K. Myrbäck (Editors), *The Enzymes* 2nd edn, Academic Press, New York, Vol. 1, pp. 305–346. [6.2, 8.6]

D.E. KOSHLAND, JR. (1959b) *J. cell. comp. Physiol.* **54**, suppl., 245–258. [6.2, 8.6]

D.E. KOSHLAND, JR., G. NÉMETHY and D. FILMER (1966) *Biochemistry* **5**, 365–385. [8.8]

D.E. KOSHLAND, JR., D.H. STRUMEYER and W.J. RAY, JR. (1962) *Brookhaven Symp. Biol.* **15**, 101–133. [9.2]

J. KVASSMAN and G. PETTERSSON (1976) *Eur. J. Biochem.* **69**, 279–287. [9.6]

K.J. LAIDLER (1955) *Can. J. Chem.* **33**, 1614–1624. [2.4]

K.J. LAIDLER (1965) *Chemical Kinetics* 2nd edn, McGraw-Hill, New York, Ch. 3. [1.7]

I. LANGMUIR (1916) *J. Amer. chem. Soc.* **38**, 2221–2295. [2.2]

I. LANGMUIR (1918) *J. Amer. chem. Soc.* **40**, 1361–1403. [2.2]

E. LAYNE (1957) *Methods Enzymol.* **3**, 447–454. [3.4]

H. LINEWEAVER and D. BURK (1934) *J. Amer. chem. Soc.* **56**, 658–666. [2.2, 2.5]

H. LINEWEAVER, D. BURK and W.E. DEMING (1934) *J. Amer. chem. Soc.* **56**, 225–230. [2.5]

O.H. LOWRY, N.J. ROSEBROUGH, A.L. FARR and R.J. RANDALL (1951) *J. biol. Chem.* **193**, 265–275. [3.4]

W.R. McCLURE (1969) *Biochemistry* **8**, 2782–2786. [3.3]

H.R. MAHLER and E.H. CORDES (1966) *Biological Chemistry* Harper & Row, New York, pp. 219–277. [6.3]

B. MANNERVIK and T. BÁRTFAI (1973) *Acta biol. med. germ.* **31**, 203–215. [10.5]

M. MARKUS, B. HESS, J.H. OTTAWAY and A. CORNISH-BOWDEN (1976) *FEBS Lett.* **63**, 225–230. [10.4]

R.G. MARTIN (1963) *J. biol. Chem.* **238**, 257–268. [8.7]

V. MASSEY, B. CURTI and H. GANTHER (1966) *J. biol. Chem.* **241**, 2347–2357. [7.7]

W. MEJBAUM-KATZENELLENBOGEN and W.H. DOBRYSZYCKA (1959) *Clin. chim. Acta* **4**, 515–522. [3.4]

J.-C. MEUNIER, J. BUC, A. NAVARRO and J. RICARD (1974) *Eur. J. Biochem.* **49**, 209–223. [8.10]

L. MICHAELIS (1926) *Hydrogen Ion Concentration* translated from 2nd German edn (1921) by W.A. Perlzweig, Baillière, Tindall & Cox, London, Vol. 1. [7.3]

L. MICHAELIS (1958) *Biograph. Mem. natn. Acad. Sci. U.S.* **31**, 282–321. [7.1]

L. MICHAELIS and H. DAVIDSOHN (1911) *Biochem. Z.* **35**, 386–412. [3.6, 7.1]

L. MICHAELIS and M.L. MENTEN (1913) *Biochem. Z.* **49**, 333–369. [2.2, 2.5, 2.8, 5.1]

L. MICHAELIS and H. PECHSTEIN (1914) *Biochem. Z.* **60**, 79–90. [2.7]

G.A. MILLIKAN (1936a) *Proc. Roy. Soc., Ser. A* **155**, 455–476. [9.3]

G.A. MILLIKAN (1936b) *Proc. Roy. Soc., Ser. B* **120**, 366–388. [9.3]

J. MONOD, J.-P. CHANGEUX and F. JACOB (1963) *J. mol. Biol.* **6**, 306–329. [8.7]

J. MONOD, J. WYMAN and J.-P. CHANGEUX (1965) *J. mol. Biol.* **12**, 88–118. [8.7]

J.M. NELSON and R.S. ANDERSON (1926) *J. biol. Chem.* **69**, 443–448. [5.3]

L.W. NICHOL, W.J.H. JACKSON and D.J. WINZOR (1967) *Biochemistry* **6**, 2449–2456. [8.9]

R. NORRIS and K. BROCKLEHURST (1976) *Biochem. J.* **159**, 245–257. [5.12, Problem 5.4]

D.B. NORTHROP (1977) in W.W. Cleland, M. O'Leary and D.B. Northrop (Editors), *Isotope Effects on Enzyme-Catalyzed Reactions* University Park Press, Baltimore, pp. 122–152. [6.8]

J.H. NORTHROP (1930) *J. gen. Physiol.* **13**, 739–766. [3.1]

C. O'SULLIVAN and F.W. TOMPSON (1890) *J. chem. Soc. (Trans.)* **57**, 834–931. [2.1, 3.6]

M.J. PARRY and D.G. WALKER (1966) *Biochem. J.* **99**, 266–274. [3.5]

A.K. PATERSON and J.R. KNOWLES (1972) *Eur. J. Biochem.* **31**, 510–517. [5.12]

L. PAULING (1935) *Proc. natn. Acad. Sci. U.S.* **21**, 186–191. [8.6]

D.D. PERRIN (1965) *Nature, Lond.* **206**, 170–171. [3.10]

D.D. PERRIN and I.G. SAYCE (1967) *Talanta* **14**, 833–842. [3.10]

M.F. PERUTZ, M.G. ROSSMANN, A.F., CULLIS, H. MUIRHEAD, G. WILL and A.C.T. NORTH (1960) *Nature, Lond.* **185**, 416–422. [8.2, 8.6]

L.C. PETERSEN and H. DEGN (1978) *Biochim. biophys. Acta* **526**, 85–92. [Problem 6.2]

G. PETTERSSON (1976) *Eur. J. Biochem.* **69**, 273–278. [9.6]

J. PIERCE and C.H. SUELTER (1977) *Analyt. Biochem.* **81**, 478–480. [3.4]

G. PORTER (1967) in S. Claesson (Editor), *Proc. 5th Nobel Symp.* Interscience, New York, pp. 469–476. [9.3]

W.R. PORTER and W.F. TRAGER (1977) *Biochem. J.* **161**, 293–302. [10.4]

B.R. RABIN (1967) *Biochem. J.* **102**, 22c–23c. [8.10]

J.G. REICH (1970) *FEBS Lett.* **9**, 245–251. [10.4]

J. RICARD, J.-C. MEUNIER and J. BUC (1974) *Eur. J. Biochem.* **49**, 195–208. [8.10]

G.R. SCHONBAUM, B. ZERNER and M.L. BENDER (1961) *J. biol. Chem.* **236**, 2930–2935. [9.2]

H.L. SEGAL, J.F. KACHMAR and P.D. BOYER (1952) *Enzymologia* **15**, 187–198. [6.3]

M.J. SELWYN (1965) *Biochim. biophys. Acta* **105**, 193–195. [3.6]

D.B. SIANO, J.W. ZYSKIND and H.J. FROMM (1975) *Arch. Biochem. Biophys.* **170**, 587–600. [10.1, 10.2]

J.R. SILVIUS, B.D. READ and R.N. McELHANEY (1978) *Science* **199**, 902–904. [7.7]

H.A. SOBER, R.W. HARTLEY, JR., W.R. CARROLL and E.A. PETERSON (1965) in H. Neurath (Editor), *The Proteins* 2nd edn, Academic Press, New York, Vol. 3, pp. 1–97. [3.4]

S.P.L. SØRENSEN (1909) *C. r. trav. Lab. Carlsberg* **8**, 1–168 (in French; a German version appeared in *Biochem. Z.* **21**, 131–304). [2.2, 7.1]

J. STEINHARDT and J.A. REYNOLDS (1969) *Multiple Equilibria in Proteins* Academic Press, New York, pp. 176–213. [7.2]

A.C. STORER and A. CORNISH-BOWDEN (1974) *Biochem. J.* **141**, 205–209. [3.3]

A.C. STORER and A. CORNISH-BOWDEN (1976a) *Biochem. J.* **159**, 1–5. [3.10]

A.C. STORER and A. CORNISH-BOWDEN (1976b) *Biochem. J.* **159**, 7–14. [8.10]

A.C. STORER and A. CORNISH-BOWDEN (1977) *Biochem. J.* **165**, 61–69. [5.7]

A.C. STORER, M.G. DARLISON and A. CORNISH-BOWDEN (1975) *Biochem. J.* **151**, 361–367. [10.1, 10.2]

J.R. SWEENY and J.R. FISHER (1968) *Biochemistry* **7**, 561–565. [6.2]

K. TAKETA and B.M. POGELL (1965) *J. biol. Chem.* **240**, 651–662. [7.3]

J.W. TEIPEL, G.M. HASS and R.L. HILL (1968) *J. biol. Chem.* **243**, 5684–5694. [5.6]

D. THUSIUS (1973) *Biochimie* **55**, 277–282. [9.5]

K.F. TIPTON and H.B.F. DIXON (1979) *Methods Enzymol.* **63**, 183–234. [7.6]

C. TSOU (1962) *Sci. Sin.* **11**, 1535–1558. [5.12]

H. VAN KLEY and S.M. HALE (1977) *Analyt. Biochem.* **81**, 485–487. [3.4]

D.D. VAN SLYKE and G.E. CULLEN (1914) *J. biol. Chem.* **19**, 141–180. [2.2]

J.H. VAN'T HOFF (1884) *Études de Dynamique Chimique* Muller, Amsterdam, pp. 114–118. [1.6]

M.V. VOLKENSTEIN and B.N. GOLDSTEIN (1966) *Biochim. biophys. Acta* **115**, 471–477. [4.2]

O. WARBURG and W. CHRISTIAN (1942) *Biochem. Z.* **310**, 384–421. [3.4]

G.B. WARREN and K.F. TIPTON (1974) *Biochem. J.* **139**, 310–320, 321–329. [6.2]

H. WATARI and Y. ISOGAI (1976) *Biochem. biophys. Res. Comm.* **69**, 15–18. [Problem 8.1]

E. WHITEHEAD (1970) *Prog. Biophys.* **21**, 321–397. [8.10]

E.P. WHITEHEAD (1978) *Biochem. J.* **171**, 501–504. [8.5]

G.N. WILKINSON (1961) *Biochem. J.* **80**, 324–332. [10.2, 10.4]

J.T. WONG (1975) *Kinetics of Enzyme Mechanisms* Academic Press, London, pp. 10–13 [2.4]; pp. 19–21 [4.1]

J.T. WONG and C.S. HANES (1962) *Can. J. Biochem. Physiol.* **40**, 763–804.
   [4.1, 6.2]
J.T. WONG and C.S. HANES (1969) *Arch. Biochem. Biophys.* **135**, 50–59. [6.9]
B. WOOLF (1929) *Biochem. J.* **23**, 472–482. [6.2]
B. WOOLF (1931) *Biochem. J.* **25**, 342–348. [6.2]
B. WOOLF (1932), cited by J.B.S. Haldane and K.G. Stern in *Allgemeine Chemie der Enzyme* Steinkopff, Dresden and Leipzig, pp. 119–120. [2.5]
A. WURTZ (1880) *C. r. hebd. Acad. Sci., Paris* **91**, 787–791. [2.1]

Solutions to problems

1.1 Order ½ with respect to A, order 1 with respect to B. An order of ½ can be rationalized by supposing the predominant species in an equilibrium mixture to be the dimer of the chemically reactive species.

1.2 (a) Slope and intercept are both inconsistent; (b) consistent; (c) slope consistent, intercept inconsistent.

1.3 That they are typically about 50 kJ mol^{-1}.

1.4 0.0033 K^{-1}.

2.1 (a) $K_m/9$; (b) $9K_m$; (c) 81.

2.3 At least 8.2.

2.4 Equilibrium constant = 5.1. The second experiment gives 1.5 for the same equilibrium constant. A change of enzyme could not by itself account for this difference. Therefore, either the reported values are unreliable or the experimental conditions were different in some unstated way (e.g. the experiment was done at a different temperature).

2.5 The more accurate methods should give values close to $K_m = 10.6$ mM, $V = 1.24$ mM min^{-1}.

2.6 (b) $V^{app} = V/(1 - K_m^A/K_s^P)$, $K_m^{app} = K_m^A(1 + a_0/K_s^P)/(1 - K_m^A/K_s^P)$; (c) When K_s^P is less than K_m^A.

3.1 At least 0.21 mM min^{-1}.

3.2 Inhibiting substances may be removed, substances added for other purposes may be activators, etc.

3.3 Selwyn plots are not even approximately superimposable, and the initial rate in (b) is about half that in (a) instead of about double as expected. The results suggest a polymerizing enzyme for which the more highly associated species is less active.

3.4 Maintain total ATP 5 mM in excess of total MgCl$_2$.

4.1 Equation 6.4 (p. 107). If B can bind to E, and the dissociation constant of EB is K_{si}^B, the second and third terms in the denominator, i.e. those in b and p respectively, must be multiplied by $(1 + b/K_{si}^B)$.

4.2 (last part) Because part of the A consumed in the reaction is converted into P′, not P.

5.1   If $K_i = (k_{-1} + k_{+2})/k_{+1}$, and if this expression is very different from
       k_{-1}/k_{+1}, then k_{-1} must be small compared with k_{+2}. So $k_{+1} \simeq 5 \times$
       $10^{-3}/10^{-4}$, or 50 M^{-1} s^{-1}. This would make k_{+1} smaller by a factor of
       about 10^4 than typical values, and suggests that the premise used for cal-
       culating it was incorrect; that is, it suggests that the original analysis by
       Kitz and Wilson (1962) was valid and that the objections raised to it by
       Childs and Bardsley (1975) were not. *See also* Cornish-Bowden (1979).

5.2   4 mM.

5.3 $K_i = 4.9$ mM, $K_i' \gg 5$ mM. The data do not allow a definite distinction
       between pure competitive and mixed inhibition. The highest s value is
       less than K_m, and so too low to provide much information about K_i'. In
       addition, the inhibitor concentrations do not extend to a high enough
       value to define either inhibition constant accurately.

5.4   (a) 1; (b) $24 \div 6 = 4$.

5.5   (a) The double-reciprocal plot; (b) ordinate; (c) as negative intercepts on
       the abscissa.

5.6   If $s < K_m$, then $K_m(1 + i/K_i)$ is larger than $s(1 + i'/K_i')$ for the same ratio
       of inhibitor concentration to inhibition constant, i.e. $i/K_i = i'/K_i'$.
       Hence with these conditions a competitive inhibitor increases the value
       of the denominator of the rate expression, i.e. it decreases the rate, more
       than an uncompetitive inhibitor. The reverse applies when $s > K_m$.
       A competitive inhibitor exerts its effect by binding to a form of the
       enzyme that predominates at low substrate concentrations, an uncompet-
       itive inhibitor by binding to a form that predominates at high substrate
       concentrations.

5.7   (a) L-Ala–L-Ala–L-Ala (because it has the largest value of k_{cat}/K_m);
       (b) Yes (cf. *Table 5.2*).

6.1   Simple bimolecular substitution reactions (designated '$S_N 2$') commonly
       proceed with inversion of configuration at the substituted atom. Reten-
       tion, as with α-amylase, can occur as a consequence of two successive
       substitutions, as in a substituted-enzyme or double-displacement mechan-
       ism. Net inversion, as with β-amylase, suggests a single substitution, as
       in a ternary-complex or single-displacement mechanism.

6.2   The mechanism as described contains no first-order steps, so the net rate
       can in principle be increased without limit by increasing the substrate
       concentrations. (In most enzyme-catalysed reactions there is at least one
       first-order step, and saturation occurs because the flux through such a
       step cannot be raised indefinitely by increasing the substrate concentra-
       tions.)

6.3   The capacity to catalyse a half-reaction is characteristic of an enzyme
       that follows a substituted-enzyme mechanism. In this case the substituted
       enzyme is likely to be a glucosylenzyme. Competitive inhibition of
       exchange by glucose indicates the occurrence of a *distinct* non-covalent
       enzyme–glucose complex: if this were the same as the glucosylenzyme the
       enzyme would be a good catalyst for hydrolysis of glucose 1-phosphate.

6.4 $v = Vab/(K_s^A K_s^B + K_s^B a + ab)$

       Plots of b/v against b would be parallel lines with slope $1/V$.

6.5   (a) Hyperbolic; (b) parallel straight lines with slope $1/V$.

6.6   Competitive inhibition for all substrate–product pairs.

6.7   Uncompetitive with respect to A; competitive with respect to B.

6.8 $\phi_0 = e_0/V$; $\phi_1 = e_0 K_m^A/V$; $\phi_2 = e_0 K_m^B/V$; $\phi_{12} = e_0 K_i^A K_m^B/V$; the point of intersection occurs at $S_1 = -\phi_{12}/\phi_2$, $S_1/v = (\phi_1 - \phi_0\phi_{12}/\phi_2)/e_0$.

6.9   (a) $K^{AC}b$; (b) none; (c) Q; (d) R.

7.1   (a) pH $= pK^{ES}$; (b) pH $= pK^E$.

7.2   Standard states are implied (cf. Section 1.3), but the definitions of the standard states have no effect on the *shapes* of the plots in *Figure 7.3*, and in particular they have no effect on the pH values at which changes in slope occur. Thus the determination of pK_a is not invalidated by the lack of proper attention to dimensions in the method.

7.3 $pK_1 = 5.99$; $pK_2 = 7.21$. The group pK values are 6.65, 6.56 and 7.10, but the specific assignments vary according to which step is assigned the value of 6.1.

7.4   Note that the rate constants for binding and release of substrate and product can only be independent of the state of protonation if H_2E, H_2ES and H_2EP have the *same* pair of dissociation constants K_1 and K_2. It is thus unnecessary to define f(h) separately for H_2E, H_2ES and H_2EP. Then,

$$K_m = \frac{k_{-1}}{k_{+1}} + \frac{k_{+2}(k_{+3} - k_{-1})\,f(h)}{(k_{+2} + k_{-2})\,f(h) + k_{+3}}$$

which can be independent of pH *either* if k_{+2} is very small *or* if $k_{+3} = k_{-1}$. In either case, $K_m = k_{-1}/k_{+1}$.

7.5   No; an Arrhenius plot of the data shows pronounced curvature, characteristic of combined effects on two or more rate constants, rather than the straight line that would be expected for the temperature dependence of an elementary rate constant.

8.1   Slope $= h - 1$. This slope has the same sign as the co-operativity.

8.2 $K_1 = \frac{1}{2}(K_1' + K_2')$; $K_2 = 2K_1'K_2'/(K_1' + K_2')$. The slope of a plot of Y against [X] must decrease monotonically as [X] is increased from zero; the plot therefore cannot be sigmoid.

8.3 $$h = 1 + \frac{(K_2 - K_1)[X]}{(1 + K_1[X])(1 + K_2[X])}$$

The extreme value of h is $2/\{1 + (K_1/K_2)^{1/2}\}$ and occurs when $[X] = (K_1 K_2)^{-1/2}$.

8.4 M_1 is an allosteric inhibitor; M_2 is an allosteric activator. (a) Increase; (b) decrease.

9.1   The simplest mechanism consistent with the results is an irreversible Michaelis–Menten mechanism, i.e. with $k_{-2} = 0$. In this case equations 9.6 and 9.7 are simplified and yield $k_{+1} \simeq 2 \times 10^6$ M^{-1} s^{-1}, $k_{+2} \simeq 60$ s^{-1}, $k_{-1} \simeq 130$ s^{-1}.

9.2   The data were calculated from $y = 16.37 + 46.3 \exp(-t/3.72) + 29.7 \exp(-t/11.41)$, but the following values would be typical of the accuracy likely with the peeling procedure described in Section 9.6: $A = 17$, $B = 42.5$, $C = 33.4$, $\tau_1 = 3.4$ ms, $\tau_2 = 10.3$ ms.

9.3   Note that the final concentrations after mixing are the same for both experiments, and the substrate concentration is high enough to compete effectively with the inhibitor. The different time constants indicate that the slower process ($1/\tau = 15$ ms) approximates to the release of inhibitor from the enzyme–inhibitor complex, i.e. $1/k_{off} \simeq 15$ ms, $k_{off} \simeq 67$ s^{-1}. Hence $k_{on} \simeq 3.3 \times 10^6$ M^{-1} s^{-1}. The observation that inhibitor release is rate-limiting in experiment (b) has no bearing on the interpretation of K_i as an equilibrium constant: this follows from the fact that it describes a dead-end reaction (*see* Section 4.4).

10.1   Least-squares: $\hat{K}_m = 1.925$ mM, $\hat{V} = 0.666$ mM min^{-1}. Distribution-free: $K_m^* = 1.718$ mM, $V^* = 0.636$ mM min^{-1}. The trend in the residuals suggests that the enzyme is subject to substrate inhibition. It would be invalid to draw any conclusions about the appropriate weights until the data have been fitted to an equation more suitable than the Michaelis–Menten equation. It would also be desirable to have considerably more than ten observations.

10.2 $\hat{K}_m = 2.747$ mM, $\hat{V} = 0.737$ mM min^{-1}; $K_m^* = 1.718$ mM, $V^* = 0.638$ mM min^{-1} (if you obtained $K_m^* = 1.677$ mM, $V^* = 0.632$ mM min^{-1}, you should read the discussion at the end of Section 10.3 on the interpretation of negative V_{ij} values). The results, when compared with those for Problem 10.1, illustrate the general point that least-squares estimates are much more sensitive than distribution-free estimates to the presence of exceptionally poor observations.

10.3 $\hat{K}_m = (\Sigma v_i^4 \, \Sigma v_i^3/s_i - \Sigma v_i^4/s_i \, \Sigma v_i^3)/(\Sigma v_i^4/s_i^2 \, \Sigma v_i^3 - \Sigma v_i^4/s_i \, \Sigma v_i^3/s_i)$;
$\hat{V} = [\Sigma v_i^4/s_i^2 \, \Sigma v_i^4 - (\Sigma v_i^4/s_i)^2]/(\Sigma v_i^4/s_i^2 \, \Sigma v_i^3 - \Sigma v_i^4/s_i \, \Sigma v_i^3/s_i)$.

10.4   In the expression for $\hat{K}_m$, both terms in the numerator have the dimensions of v^7/s; both terms in the denominator have the dimensions of v^7/s^2; thus the expression as a whole has the dimensions of s, which is correct. Similarly, the expression for $\hat{V}$ has the dimensions of v^8/s^2 divided by v^7/s^2, or v, which is also correct.

Index